Alexander Tweedie

Lectures on the Distinctive Characters, Pathology and Treatment

Of Continued Fevers, Delivered at the Royal College of Physicians of London

Alexander Tweedie

Lectures on the Distinctive Characters, Pathology and Treatment
Of Continued Fevers, Delivered at the Royal College of Physicians of London

ISBN/EAN: 9783337162955

Printed in Europe, USA, Canada, Australia, Japan

Cover: Foto ©berggeist007 / pixelio.de

More available books at **www.hansebooks.com**

LECTURES

ON THE

Distinctive Characters, Pathology, and Treatment

OF

CONTINUED FEVERS,

DELIVERED AT THE

ROYAL COLLEGE OF PHYSICIANS OF LONDON.

BY

ALEXANDER TWEEDIE, M.D. F.R.S.

FELLOW OF THE ROYAL COLLEGE OF PHYSICIANS OF LONDON;
CONSULTING-PHYSICIAN TO THE LONDON FEVER HOSPITAL;
PHYSICIAN TO THE FOUNDLING HOSPITAL;
EXAMINER IN MEDICINE IN THE UNIVERSITY OF LONDON; ETC. ETC.

LONDON:

JOHN CHURCHILL, NEW BURLINGTON STREET.

MDCCCLXII.

In the following Lectures on the various types of Continued Fevers, I have restricted myself as much as possible to practical details, avoiding discussions on their proximate cause—a subject of which so little is known, and on which the most discordant views have been entertained.

I have dwelt, tediously perhaps, but earnestly, on the importance of discriminating the forms which fevers assume, and more especially on the phenomena by which enteric (typhoid) is distinguished from typhus fever; and I hope the arguments I have advanced, corroborated by the opinion of others who have investigated the subject carefully and dispassionately, may tend to convince such of my professional brethren as are still in doubt on this essential part of the pathology of fevers.

Pall Mall,
  September, 1862.

CONTENTS.

PLATE I.

In this drawing the rose-coloured spotted eruption peculiar to enteric (typhoid fever) is shown. Each spot is distinct, of a circular shape, varying in size from a point to a line and half—rarely exceeding two lines—in diameter. They are most generally discovered on the chest and abdomen, but may occasionally be detected on the extremities also.

PLATE II.

Represents the mulberry eruption diagnostic of typhus fever. It exhibits the irregular outline of each spot, which is in size from a point to three or four lines in diameter, the larger spots being formed by the coalescence of the smaller. The congeries of these larger spots in on the trunk and extremities, gives the skin a diffused mottled appearance.

PLATE III.

Shows the lesions which take place in the lower portion of the ileum : — ulceration of Peyer's patches and the changes in the solitary glands. The enlargement of the glands of the mesentery is also represented.

In Peyer's patches—the number and size of the ulcers increasing as they approach the ileo-cæcal valve—their varying shape or form, some being round, others elliptical, or more or less irregular, are indicated.

Of the solitary glands, some are represented simply enlarged ; in others, ulceration has taken place, the shape of the ulcer being round, not oval or elliptical.

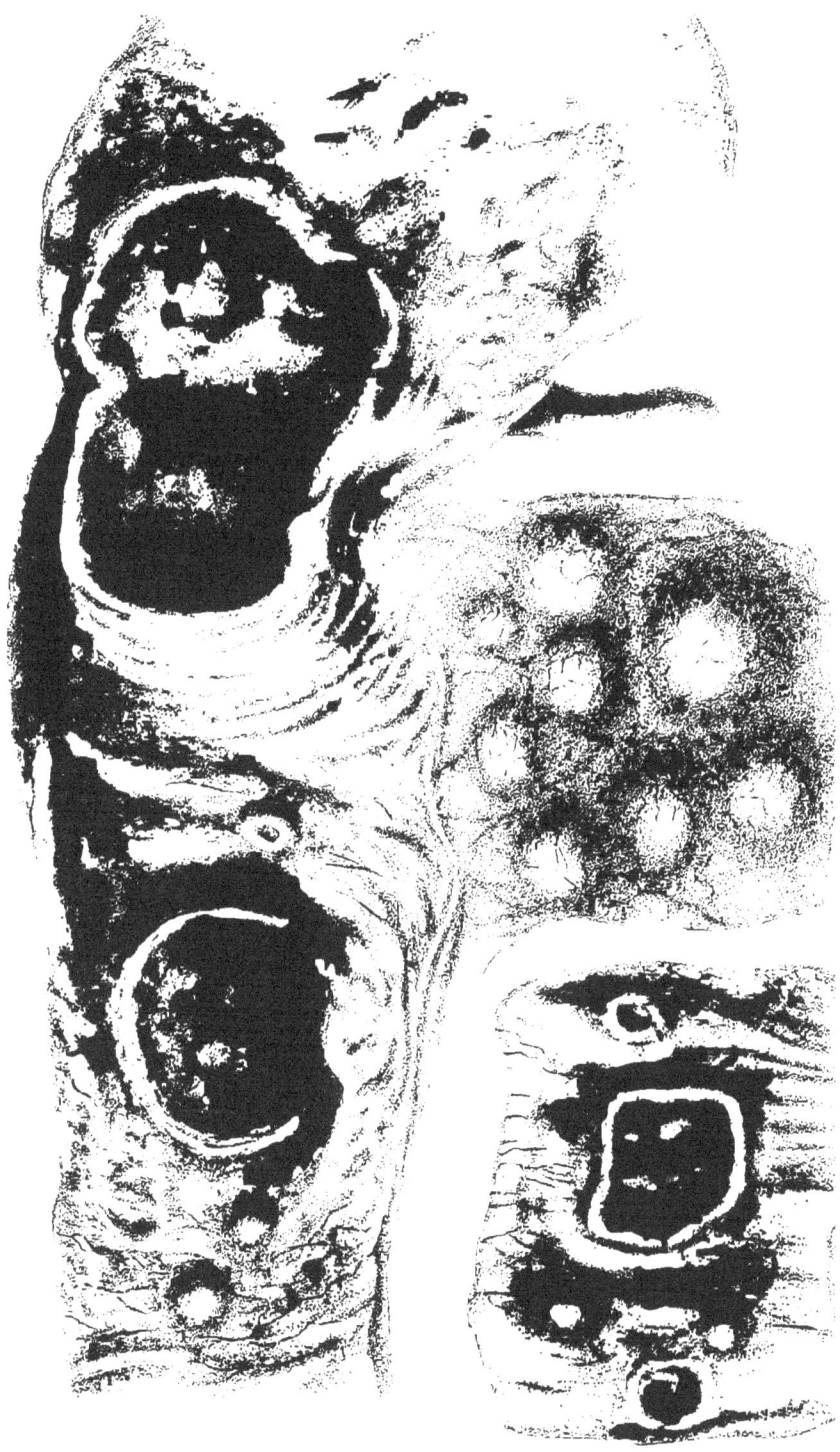

PLATE IV.

Illustrates intestinal perforation in the centre of an ulcerated patch. The drawing was taken from a preparation exhibiting this lesion. It occurred in a youth seventeen years of age, in the third week of enteric fever of moderate severity.

LECTURES ON FEVERS.

MR. PRESIDENT,

When I was requested by our late distinguished and lamented President,* shortly before his decease, to deliver the Lumleian Lectures in this theatre, he suggested that the subject should be the Pathology and Treatment of Fevers, as they are presented to our observation in this country. I admit, that for many reasons, I had great difficulty in accepting the nomination. I felt, on the one hand, that the audience whom I should address comprised the most distinguished members of our profession—many of them hospital physicians of great and deserved reputation—who have enjoyed equally favourable opportunities as myself of becoming practically acquainted with the complex group of maladies which I should have to discuss.

* Dr. Paris.

B

But, on the other, I could not overlook the circumstance, that in the ordinary professional intercourse
with my medical friends, I am almost daily interrogated as to my views of the pathology of fevers,
and more especially as to the modes of treatment,
in many respects widely dissimilar, which have
been more recently proposed.

I must also in candour confess, that another and
more personal consideration presented itself to my
mind—viz., that of late years, the opinions I formerly entertained on some of the cardinal points
have undergone considerable change, and I only
waited for a fitting opportunity to bring my more
matured ideas before the profession.

I therefore resolved to accept the offer, as affording me the most desirable means of discussing the
general subject of the pathology of fevers as they
occur in Britain.

Let me, however, remark, that the more important points having recently undergone minute investigation, more minute and searching, perhaps,
than that of any other class of acute diseases, the
subject must be already well known to those of my
audience who have kept their knowledge in advance. To such of my professional brethren,
therefore, I shall have little to bring forward that
is novel, or with which they are not already acquainted. But there may be others to whom,
from want of leisure or of opportunity, much of
what has been recently advanced is less intimately

known. I shall therefore endeavour in this course of lectures to give a condensed, yet, I trust, faithful account of the present state of our knowledge on this important class of maladies.

It is generally admitted that few diseases are more replete with interest than fevers, whether we consider the epidemic visitations which, in times more or less remote, have appeared in different quarters of the globe, the peculiar or characteristic features of their varied types or forms, but, above all, the great sacrifice of life which they have often occasioned.

Take for example the Plague, which, within the last two centuries, spread with fearful rapidity over Europe, but happily, having now ceased to exist, remains only a matter of history. It is recorded that in three years in the fourteenth century (1347 to 1350), during which period this pestilence raged in Europe, no fewer than 25,000,000 human beings —nearly a fourth of the entire European population—were destroyed by it; and when the disease broke out in this metropolis on several occasions in the sixteenth and seventeenth centuries, more than 150,000 of the inhabitants perished, the hand of death sparing neither rich nor poor. Again we are told that of the plague in 1593 (between the months of March and December) 11,166 died; in that of 1603, the deaths were nearly 30,000; in that of 1625, the mortality was 34,734; in that of 1636, (distinguished by the name of the Great Plague, from its lasting twelve years,) upwards of 11,000 persons fell victims in the short space of eight months; while in that of 1665 — the year preceding the Great Fire of London, and happily the last appearance of the pestilence in this metropolis —the mortality within twelve months was little short of 70,000.

Without pursuing the history of the plague in other parts of Europe, I may incidentally allude to its appearance in Moscow in 1725, in which year, out of a population of 90,000, at least 40,000 persons were destroyed. It reappeared in that city in the year 1771, and carried off 80,000 of the inhabitants.

Let me allude to another pestilence which appeared in the fifteenth and sixteenth centuries in this and other parts of Europe, to which the name of sweating fever or sweating sickness was given, from a prominent symptom with which it was attended—viz., a profuse, often fœtid, perspiration. It is described by Hecker as a violent inflammatory fever, ushered in by a short rigor, followed by sudden perspiration, præcordial oppression, headache, and lethargic stupor, and terminating by a profuse sweat. The whole phenomena of the disease were of a few hours' duration, seldom exceeding thirty-six hours. We are told that the sense of internal heat was intolerable, yet that any refrigerant was certain death. It was observed, too, of this singular malady, that its principal feature — the colliquative sweat — was sometimes absent, just as in the more dangerous forms of cholera there may be neither vomiting nor purging. In other cases, lividity of the countenance, urgent dyspnœa, violent vomiting, or, occasionally, convulsions, were superadded to the other characteristic features of this strange disease. It is no wonder, then, that it occasioned general consternation, not more from its suddenness, violence, and rapidly fatal termination, than from the numbers attacked, scarcely one in a hundred, as we are told, escaping.

This sweating fever was first observed in the more densely-crowded districts of London in the autumn of 1485. So sudden and rapid was its progress, that it was no uncommon circumstance for persons who retired in perfect health at night, to be in the morning numbered with the dead. Even when an individual was so fortunate as to escape death, there was no security against its recurrence, many being seized, and with equal violence, a second, and, it might be, a third time. Though it lasted about five weeks only on this its first appearance, many thousands are reported to have died of it. Nor was this scourge confined to the metropolis; for it is recorded, that towards the end of the same year, it appeared in many parts of England, exhibiting the same virulence, and after raging for a few weeks, suddenly ceased without leaving a trace of its existence.

It reappeared, however, early in the following century, at four successive periods, (in 1506, 1517, 1528-29, and 1551,) sometimes with greater, sometimes with less intensity, than on its first eruption. In 1551 we find that it broke out at Shrewsbury, whence it spread from place to place, till at length it again reached London, causing from its frightful mortality even greater alarm than on its first outbreak. The total number of its victims was not estimated, but it must have been very great, as Godwin has alluded to it as a depopulation. It finally vanished in the autumn of that year, and happily has never since been known in Europe, or any quarter of the globe.

The resemblance of this appalling disease in some of its features to an occasional epidemic of our own times—the relapsing or recurrent fever, to be afterwards brought under notice—is striking. The suddenness of invasion, its comparatively short duration, and abrupt termination by profuse critical sweat, and in many instances, after convalescence, a relapse of the same symptoms, ending again by critical perspiration, are all circumstances showing strong points of resemblance; though there is one very important difference, that the one disease—the sweating fever—was wide-spread and mortal, the other is generally limited, and seldom fatal.

I may also for a moment allude to another pestilence, happily unknown in this country—the yellow fever—which has at times committed great

ravages on the western shores of America, in the
West Indies, the eastern coast of Spanish America,
and in Africa. Its outbreak has generally been
traced to intense solar heat, acting more especially
on marshy districts, showing that a certain elevation of temperature is required for its production.
It is accordingly met with in localities between 40°
north and 20° south, where the summer heat
ranges from 75° to 80°, or more; and it is stated
by those who have watched its progress, that a
reduction of atmospheric temperature has a greater
effect in checking its extension than storms of wind
or rain, a marked decline being often suddenly observed when the temperature becomes lower.

The mortality from this form of fever is sometimes alarming. When the disease first appears
few indeed of those attacked recover; but as it
becomes more prevalent among the population, the
death-rate diminishes in proportion to the numbers
that are seized.

The epidemics which have appeared within
the memory of many of us, and even long before,
have been mercifully stripped of the terrors of those
to which I have thus briefly alluded. Our modern
epidemics (with the exception of cholera and in-
fluenza) have been restricted to some form of continued fever, prevailing more or less extensively in
different places; never entirely absent, but having
an apparent connexion with overcrowding, de-

fective drainage, insufficient water supply, but
more especially scarcity and famine.

In no part of the British dominions has fever
existed to such an extent as in Ireland, faithful
accounts of which have been published in the
excellent reports of Drs. Barker and Cheyne,
from which it would appear, that the Irish epidemics,
in respect of the numbers affected and great mortality, resembled the plague of the sixteenth and
seventeenth centuries. It has been computed that
by the fever that raged throughout Ireland in the
middle of last century (1740–1741) upwards of
80,000 were carried off.

Subsequent to the period to which I allude, there
have been several epidemic visitations, and of
which records have been preserved, affording an
opportunity of comparison with previous and subsequent outbreaks. For example, Dr. Rutty, in
1741, specially alludes to a fever of six or seven
days' duration, terminating in a critical sweat, the
patient being subject to a relapse even to a third
or fourth time, which is a short but accurate description of relapsing fever.

The most serious epidemic, however, with which
Ireland has been visited was that of 1816-17.
It evidently originated in the general distress,
combined with the scarcity, bad quality, and high
price of provisions; and from this apparent origin
it received the name of the famine fever. Though
fever had been smouldering in many districts for

some time previously, it broke out with great violence about the beginning of 1817, and it was computed that between the early part of 1817 and the middle of 1819, embracing a period of eighteen months only, out of a population of six millions, more than a million and a half were attacked. In Dublin alone, in little more than two years, upwards of 42,000 persons—about a fifth of the whole population—were seized. It was, moreover, ascertained from returns transmitted to Government, that in the two years mentioned at least 65,000 fell victims to epidemic fever in Ireland.

Happily this wide-spread malady gradually abated; and though fever has not entirely ceased throughout the sister island, and probably never will, its prevalence and mortality have greatly abated within the last forty years. The establishment of new colonies in various quarters of the globe, by relieving Ireland, amongst other places, of her unemployed and starving population, and the prosperous results of the Encumbered Estates Act, by which an immense stimulus has been given to industrial habits, to better modes of cultivation, and consequently to increased comforts amongst the peasantry, have doubtless tended greatly to check the prevalence of fever.

From a return I have recently obtained from the House of Recovery and Fever Hospital in Dublin, it appears that between the years 1820 and 1857, the total number of fever cases admitted was

139,611; the annual average being 3771. The largest number admitted in a single year (1827) was 10,882, and the smallest (1857) 1596. The population of Dublin at the last census was about 258,369.

If the condition of England and Scotland in reference to the prevalence of epidemic fever since the beginning of the present century be contrasted with that of Ireland, the comparative immunity we have enjoyed on this side the Channel is striking. On referring to the register of patients admitted into the London Fever Hospital, I find that since its establishment in 1802 until the end of 1857 the total numbers admitted have been 24,763. It is proper to state, however, that in this summary are included other acute diseases sent into the hospital with certificates of fever, but which, on investigation, were discovered not to be cases of fever, but some form of local inflammation of which the fever was merely symptomatic. The exact proportion which such acute diseases bear to the cases of fever is therefore conjectural. It is only within the last ten years that the cases have been classified in the hospital register. I may also mention that the Fever Hospital was at first established on a very small, or, I may say, experimental scale, and has only within the last few years been reconstructed and extended to its present more useful size, so as to be capable of

accommodating two hundred patients. It is also
to be kept in view, in estimating the prevalence of
fever in and around this metropolis, that fever
cases are received in limited numbers into the
general hospitals; while in many parishes or unions,
contrary to the injunctions of the Poor-law Board,
patients with fever are treated in the infirmaries
attached, until the symptoms, in many instances,
assume a formidable, and too often hopeless aspect,
when they are conveyed to our spacious wards—
sometimes in a dying state—only to swell our
hospital mortality. Many of the sick poor, also,
when seized with fever, are attended at their own
abodes by the medical officers attached to the
numerous dispensaries with which our metropolis
abounds; and I willingly embrace the present
opportunity of offering a tribute of praise to those
members of our profession, who, with a self-sacrifice
too often unknown and unrequited, give their
services in a truly Christian spirit, regardless of
the danger to which, in their sense of duty, they
expose themselves.

But taking into consideration the rapidly increasing population throughout England, and the
consequent manifold predisposing circumstances,
there is great cause of thankfulness that visitations of fever have been comparatively so rare,
compared with the Irish epidemics adverted to.

In regard to Scotland, if we take as data the
records of the infirmaries and Fever Hospitals of

Edinburgh and Glasgow, there seem to be many more predisposing and exciting causes of fever in activity in the northern parts of the kingdom, than in London at all events, if not in the other large towns in England. From a statement recently published by Dr. Christison of the fever cases treated in the Edinburgh hospitals since the commencement of the present century, the total numbers were 45,189; from which it appears that, for a population under 200,000, the numbers are considerable, much larger, relatively, than in London or even Glasgow, but infinitely smaller in proportion to the population, than in any of the provinces or districts of Ireland during the pestilential years. In three years only (1828, 1838, and 1848) did the number of fever patients received into the Edinburgh hospitals reach 2000, the largest being in 1848, when 4693 were admitted. It is, however, to be kept in view that the admissions into the hospitals in Edinburgh, as in this metropolis, give only a comparative idea of the prevalence of fever, as a very considerable number of cases, much larger than in London, are visited at their own dwellings, under the admirable arrangements in the northern metropolis for relieving the wants of the poor when visited by sickness.

From the statistical account of fever in Glasgow published by Dr. Cowan, that city appears to have been peculiarly liable to epidemic visitations. Combining the numbers treated in the hospitals with the patients visited at home by district medical officers duly appointed, Dr. Cowan states, that in ten years (from 1828 to 1837), for a population of about 200,000, there had been 28,290 cases of fever, the smallest annual number being 1089, the largest 7707. Glasgow, therefore, of all the larger towns in Britain, was, for some reasons, in the years referred to, the most unhealthy as regards epidemic fever.

In the principal towns in England, the favourable contrast is striking. From the same report, we find that, while in Glasgow the average number of fever cases for seven years prior to 1837 was 1842 annually, in Manchester, with a population at the same period of 228,000, the yearly number was only 497; in Leeds, with a population of 123,000, only 274 ; and in Newcastle, with a population of 58,000, it was not more than 39.

Another striking fact is brought out, that while the average number of fever patients under treatment in the Glasgow hospitals, between 1797 and 1806, did not exceed 88 annually, it had increased on an average of seven years (1831 to 1837) to 1842; but in Manchester, notwithstanding the rapid increase of population, the average had scarcely varied, being for the early period (1797 to 1806) 462, and for the seven years prior to 1837 scarcely in excess—viz., 497 only.

From communications I have received lately from Glasgow, Manchester, and Leeds, I find that in Glasgow the prevalence of epidemic fever continued from 1837 to 1847, but that for the last ten years there has been a sensible diminution, the average annual admissions into the Infirmary scarcely exceeding 800.

In Manchester, again, with the exception of one violent outbreak in 1847—evidently the consequence of a sudden large importation of Irish immigrants — the same comparative immunity from epidemic fever has obtained. Dr. Noble states, in a communication with which he has favoured me, that since that year (1847) there has been no epidemic fever, and, in consequence, the hospital for fever patients has been given up, one ward in the general hospital being appropriated to the few cases now and then applying for admission. A few patients also are accommodated by the guardians of the poor, but he adds, of late any true fever is quite a rarity in Manchester.

In Leeds, the same diminution of fever cases is reported. Since the period referred to in Dr. Cowan's statement, the average annual admissions into the House of Recovery have not reached 400, for a population now upwards of 172,000. Fever patients are not received into the Leeds General Infirmary.

These statements are not without importance, more especially in reference to the investigation of the origin and propagation of epidemic fevers in those places where they appear to be localized, as well as to the circumstances which apparently tend to diminish such local susceptibility.

I recommend this subject, so pregnant with interest, and in which every class of the community is deeply concerned, to the patient investigation of those who have the opportunity and inclination to cultivate the etiology of diseases. From the zeal recently displayed by my colleague Dr. Murchison, in tracing the probable sources of the several forms of fever, I anticipate from him the enunciation of some important facts touching their origin and mode of propagation.

CLASSIFICATION OF CONTINUED FEVERS.

After these cursory observations on epidemics, I proceed to bring before you a description of the several forms of continued fever which prevail in this and other temperate climates, with such modifications in the symptoms as have come under my individual observation in hospital and private practice; and I may remark, that from my connexion with an hospital specially appropriated to this class of diseases, I am enabled to lay before you a statistical view of their forms, their prevalence at particular periods or seasons, as well as various pathological facts, all tending to invalidate the doctrine formerly taught, and by many still maintained, of the identity of the several varieties of continued fever. I need scarcely remind you that, until a comparatively recent period, continued fevers were all included in one class, under the generic term typhus—a designation manifestly inapplicable to several varieties (or I may even say diseases) differing from each other in their origin, anatomical characters, and progress —in short, in all essential particulars; the term itself (typhus) being derived from a single symptom not always present, and having the sanction of antiquity only for its general application. Still, it distinguishes, and should be restricted to one form, characterized by early prostration, more or less prominent affection of the nervous system, but more especially by the invariable absence of any specific lesion, as will be presently pointed out.

Continued fevers may be classified under four distinct varieties: 1. Typhus. 2. Enteric (intestinal or typhoid) fever. 3. Relapsing fever. 4. Febricula.

Let me define each of those different forms.

1. Typhus fever originates in circumstances tending to impair the essential or vital properties of the blood, more especially overcrowding, defective ventilation, insufficient nourishment, and hence its prevalence in times of scarcity and famine. Its accession is marked by no special symptoms, but such as occur in many acute diseases—chilliness alternating with heat of skin; quickened pulse, succeeded by muscular prostration ; more or less sensorial disturbance; and between the fifth and eighth day a peculiar morbillous-like eruption, not fading on pressure, and persistent; the duration of

the fever being about fourteen, seldom exceeding twenty-one days. In fatal cases there is no
specific lesion, congestion of the internal organs
being the only change observed. If there be other
lesions, they are superadded or accidental.

2. Enteric or intestinal fever.—This form,
(known also by the term typhoid,) is endemic, and
supposed to be produced by emanations from
organic matter. Its mode of invasion, which is
slow and insidious, differs little from that of the
other forms, except that there is almost invariably
diarrhœa from the commencement, followed often
by gurgling in the right iliac fossa, and more or
less tympanitic distension. It is characterized,
moreover, by an eruption of rose-coloured spots,
visible about the eighth day—often later—first on
the anterior aspect of the trunk, rarely on the face
or extremities, coming out in successive crops, and
fading or entirely disappearing on pressure ; occasionally epistaxis, little comparative diminution
of strength, and sometimes sensorial disturbance.
This form is often protracted, seldom terminating
before the third week, often lasting much longer.
After death there is invariably alteration in the
solitary and agminated glands of the ileum, which
are enlarged and more or less extensively ulcerated,
according to the duration of the fever, with enlargement and softening of the corresponding
mesenteric glands, and increased volume and softening of the spleen.

C

3. Relapsing fever.—In this peculiar form the invasion is sudden, marked by irregular chills or prolonged shivering, succeeded by hot skin, severe pain in the head and limbs, epigastric tenderness, sometimes vomiting (often accompanied with jaundice), and hepatic and splenic congestion. But the chief peculiarity consists in an abrupt cessation of the symptoms from the fifth to the seventh day, so that the patient is apparently restored to convalescence. A few days afterwards, however, generally about fourteen days from the beginning of the attack, a well-marked relapse, or recurrence of the symptoms takes place, which, after a profuse sweat, about the third day again disappear, leaving the patient exhausted, though soon recruited by a long invigorating sleep. This form is rarely fatal, and leaves no special lesion. It appears to be connected with famine and destitution.

4. Febricula.—This is a mild variety of fever, characterized by its short duration and the mildness of the symptoms. It seldom lasts more than a few days; sometimes terminating within twenty-four hours, when it constitutes what has been called ephemera, or one-day fever. Its cessation is generally preceded by a critical sweat. It is seldom if ever fatal, unless from the supervention of acute local disease, which of itself may destroy life.

Now let me remark that these are not arbitrary, but practical distinctions, recognised in hospital as well as in private practice, and occurring either sporadically, or as epidemic visitations.

The records of the London Fever Hospital afford the most satisfactory mode of showing the comparative prevalence of each of these forms in and around this metropolis, and of these my colleague Dr. Murchison has availed himself, in his endeavours to trace their etiology and mode of propagation. The details required no small amount of labour; and I can safely state that, being collated by him, their accuracy can be relied on. Each case included in this statistical record was carefully noted, the diagnosis being entered at its termination in a separate column in the hospital register.

The numbers of each form admitted during the last ten years have been arranged in tables, so that at a glance the total of each for every year is exhibited.

TABLE I.

Admissions into the London Fever Hospital from 1848 *to* 1857, *inclusive.*

Forms of Fever.	1848.	1849.	1850.	1851.	1852.	1853.	1854.	1855.	1856.	1857.	Total.
Typhus	526	155	130	68	204	408	337	342	1062	274	3506
Enteric	152	138	137	234	140	211	228	217	149	214	1820
Relapsing	13	29	32	256	88	16	5	1	1	0	441
Febricula	16	79	62	6	129	152	144	62	89	72	861
Total for each year	707	401	361	614	561	787	717	662	1300	561	6628

This table shows, that of 6628 cases admitted,

c 2

there were, of typhus, 3506; of enteric or intestinal fever, (typhoid,) 1820; of relapsing or recurrent, 441; and of febricula, 861. The preponderance of typhus proves, therefore, that it is the endemic fever of the metropolis, and, very probably, of Great Britain and Ireland. It also appears that not only was the number of cases of typhus nearly double those of enteric fever, but that the proportion of the one to the other varied greatly in different years. Thus, in 1848, there were 526 of typhus to 152 of enteric fever; in 1849, 155 to 138; in 1850, 130 to 137, (showing in that year a slight preponderance of enteric fever;) in 1851, 68 to 234, (being a greatly larger increase of enteric fever, and the year when there was the greatest number of cases of relapsing fever;) in 1852, the proportion was as 204 to 140; in 1853, as 408 to 211; in 1854, as 337 to 228; in 1855, as 342 to 217; in 1856, as 1062 to 149; in 1857, as 274 to 214.

Therefore, in every year, with two exceptions, (those of 1850 and 1851,) typhus was the predominant form. There has, however, throughout the whole period of ten years, been less annual variation in the prevalence of enteric fever than of typhus, the highest number of typhus being 1062, and the lowest 65; while of enteric fever, the highest was 234, and the lowest 137.

It is also brought out from this table, that when fever prevails as an epidemic, the form it assumes is that of typhus, enteric fever being apparently

more under the influence of some hitherto unexplained local causes, or connected with particular
seasons of the year.

In order to show the influence of months and
seasons on the prevalence of fever, the table No. II.
has been constructed.

TABLE II.

*Table showing the Number of Admissions of the different Forms of Con-
tinued Fever into the London Fever Hospital for Ten Years, in each
Month and Quarter.*

ADMISSIONS MONTHLY.

	Typhus.	Enteric Typhus.	Relapsing.	Febricula.	Total.
January	385	113	22	68	588
February	300	85	27	67	479
March.................	389	77	17	69	552
April	380	60	46	69	555
May.......................	396	79	46	57	578
June	312	119	43	75	549
July.......................	280	157	31	75	543
August	239	233	44	81	597
September	206	260	23	76	565
October	214	253	56	77	600
November	211	223	49	80	563
December	194	161	37	67	459
Total	3506	1820	441	861	6628

ADMISSIONS QUARTERLY.

First Quarter	1074	275	66	204	1619
Second ,, ...	1088	258	135	201	1682
Third ,, ...	725	650	98	232	1705
Fourth ,, ...	619	637	142	224	1622
Total	3506	1820	441	861	6628

It shows, first, as regards typhus, that the
largest numbers admitted were, in March, (389,)
April, (380,) and May, (396;) the numbers de-
creasing in the last four months of the year. Second, that enteric fever appears to be most prevalent in September, (260,) October, (253,) and

November, (223,) apparently increasing in the
autumn months, in which typhus becomes less
common.

There seems also a great variation in the two
forms in the different seasons. Thus Table No. II.
shows the following results in each quarter or
season :—

	Typhus.	Enteric fever.
January, February, March	1074	275
April, May, June	1088	258
July, August, September	725	650
October, November, December	619	637

It seems, therefore, that typhus is most prevalent in spring, and the least so in autumn; while
enteric fever is least prevalent in spring, and most
prevalent in autumn.

Third. Relapsing fever, which is only an occasional visitant, is not apparently under any particular influence as to months or seasons. It
showed an epidemic character in 1851, when 256
cases were admitted, and it was probably confined
to the poorer classes, as I do not remember to have
met with a single case in private practice. The
next year (1852) showed a considerable diminution
(88 only); in the following year there were 8 cases;
and in 1854, 5. It has not been since observed,
with single exceptional instances.

Fourth. Neither does febricula—that mild non-
descript form of fever—seem to be in any degree

influenced by season, the numbers being pretty uniform throughout the year.

As to the mean age and the sex of each form— by the results of Table III. we find that age appears to exert decided influence on the liability to fever. The mean age of all the forms is shown to be nearly 25·96; of typhus, 29·33; of enteric fever, 21·25; of relapsing fever, 24·41; and of febricula, 22·82.

The same table shows the little influence sex has on either of the two forms. Taking the experience of ten years, of 6628 there were 3324 males and 3304 females, showing an excess in favour of males of 20 only.

TABLE III.

Table showing the Total Number of Cases, the Mean Age, and the Sex of each of the Forms of Continued Fever admitted into the London Fever Hospital during a period of Ten Years.

MEAN AGE AND SEX.

Forms of Fever.	Admissions.	Mean Age.	Sex.	
			Males.	Females.
Typhus	3506	29·33	1737	1769
Enteric	1820	21·25	905	915
Relapsing	441	24·41	233	208
Febricula	861	22·82	449	412
Total	6628	25·96	3324	3304

In regard to the number of admissions and number of deaths of each of the forms for each quinquennial period of life, we find from Table IV., which exhibits the ages of each of the forms for each quinquennial period, that the period of life at which continued fever of every form most commonly occurs is from 15 to 20; the second from 20 to 25; the third, from 10 to 15; and the fourth, from 25 to 30. After the age of 40, typhus greatly preponderates, and at the extremes of life —infancy and old age—it is very common.

TABLE IV.

Table showing the Number of Admissions and the Number of Deaths of each of the Forms of Continued Fever for each Quinquennial Period of Life.

AGES.	TYPHUS. Admissions.	Deaths.	ENTERIC. Admissions.	Deaths.	RELAPSING. Admissions.	Deaths.	FEBRICULA. Admissions.	Deaths.
Under 5 years	17	3	4	0	4	0	8	0
From 5 to 10 yrs....	183	14	103	14	32	0	83	0
,, 10 — 15 ,, ...	363	18	250	32	63	1	133	0
,, 15 — 20 ,, ...	546	26	519	84	92	1	186	0
,, 20 — 25 ,, ...	595	48	404	81	76	0	163	0
,, 25 — 30 ,, ...	343	52	240	45	37	0	60	0
,, 30 — 35 ,, ...	323	55	100	30	37	3	67	0
,, 35 — 40 ,, ...	270	89	60	14	19	2	45	0
,, 40 — 45 ,, ...	292	87	46	8	40	1	44	0
,, 45 — 50 ,, ...	212	82	20	5	8	2	20	0
,, 50 — 55 ,, ...	150	77	8	2	15	0	12	0
,, 55 — 60 ,, ...	100	51	9	4	7	1	9	0
,, 60 — 65 ,, ...	88	49	7	4	5	0	6	0
,, 65 — 70 ,, ..	42	28	1	1	1	0	5	0
,, 70 — 75 ,, ...	24	17	...	...	1	0	3	0
,, 75 — 80 ,, ...	6	5	1	...	...	...	1	0
Above 80 years	2	2	...	...	...	...	...	0
Not ascertained	50	11	48	5	4	...	16	0
Total	3506	714	1820	329	441	11	861	

If we run the eye along the typhus column, we find the following results :—

Of 3506 cases, there were—

Under 10 years of age 200

Between 10 and 15 years of age . 363

  ,,   15 and 25   ,,   . . 1041

  ,,   25 and 35   ,,   . . 666

  ,,   35 and 45   ,,   . . 562

  ,,   45 and 55   . . 362

Between 55 and 65	,,	. .	188
,, 65 and 75	,,	. .	66
,, 75 and 80	,,	. .	6
Above 80	,,	. .	2
Ages not ascertained		. . . .	50

Total of typhus fever . . 3506

As examples of extreme ages, there was one of an infant six months old, and of a male aged 84 (both spotted).

If we contrast the column for enteric fever, we find it agree with that for typhus, in so far as it shows that the age at which enteric typhus most frequently occurs is from 15 to 20; the next, from 20 to 25; the third, from 10 to 15; and the fourth from 25 to 30. The average proportion of cases, however, between 15 and 25 (of enteric fever) exceeds that of typhus by more than one half, or in the ratio of 923 to 1820.

Of the 1820 cases of enteric fever, there were—

Under 10 years of age		. . . .	107
Between 10 and 15 years of age		.	250
,, 15 and 25	,,	. .	923
,, 25 and 35	,,	. .	340
,, 35 and 45	,,	. .	106
,, 45 and 55	,,	. .	28
,, 55 and 65	,,	. .	16
,, 65 and 70	,,	. .	1
,, 75 and 80	,,	. .	1
Ages not ascertained .		. . .	48

Total of enteric fever . . . 1820

Extreme youth appears to be little less under the influence of enteric fever than typhus. After the age of 45, there is a rapid decrease in the numbers, and above the age of 50, the proportion is about one per cent. only; while in typhus, the proportion above 50 years of age, is about 12 per cent. In short, the susceptibility to enteric fever diminishes in the middle and advanced periods of life. It selects manhood in preference to either extreme.

The result of the column of relapsing fever tends to show that it is most prevalent between the ages of 15 and 20; next, between 20 and 25; and, thirdly, between 10 and 15.

Of the 441 cases there were—

Under 10 years of age	. . . .	36
Between 10 and 15 years of age	.	63
„ 15 and 25 „	. .	168
„ 25 and 35 „	. .	74
„ 35 and 45 „	. .	59
„ 45 and 55 „	. .	23
„ 55 and 65 „	. .	12
„ 65 and 70 „	. .	1
„ 70 and 75 „	. .	1
Ages not ascertained	. . . .	4
		441

Young subjects seem to be more susceptible of his form than of even enteric fever. Age, however, on the whole, lessens the susceptibility to a greater degree than to typhus, but to a much less

degree than to enteric fever. The youngest subject
of this form was a female child two years of age,
and the oldest a male aged 74.

The column for febricula indicates that a much
larger proportion of this form occurred in extreme
youth than in any of the others, more than ten per
cent. of the cases being under 10 years of age.
From the number of cases above 30, amounting to
about $26\frac{1}{2}$ per cent., it appears that advanced life
is less susceptible to febricula than to enteric fever,
and less so than even to the relapsing fever.

Of the 861 cases, there were—

Under 10 years of age		91
Between 10 and 15 years of age .		133
,, 15 and 25 ,, . .		349
,, 25 and 35 ,, . .		127
,, 35 and 45 ,, . .		89
,, 45 and 55 ,, . .		32
,, 55 and 65 ,, . .		15
,, 65 and 75 ,, . .		8
,, 75 and 80 ,, . .		1
Ages not ascertained		16
		861

For these statistical details I am indebted to
Dr. Murchison, who has devoted much time and
labour on a department of the pathology of fever
of extreme interest and value.*

---

* Since the present course of lectures was delivered, I have ascertained the number of patients admitted, with the different forms of

Having laid before you these statistical details with the view of showing the prevalence of the different types of fevers, I am desirous to bring under your notice a question which has of late years been agitated and keenly discussed — the identity or non-identity of the two principal forms, the typhus and enteric (or typhoid). I prefer the term enteric, and shall employ it in future.

Before, however, going into the arguments bear-

fever, into the London Fever Hospital for the last four years and a-half. The following table gives the classified results :—

Admitted in the Year.	1858.	Died.	1859.	Died.	1860.	Died.	1861.	Died.	From Jan. 1 to June 30, 1862.	
With typhus fever .	15	9	48	18	25	10	86	15	1098	226
,, enteric fever .	180	26	176	32	94	26	161	32	67	12
,, febricula . .	44	0	34	0	32	0	49	0	55	0
	239	35	258	50	151	36	296	47	1220	238

From the above statement it appears, that in the years 1858, 1859, and 1860, the proportion of cases of enteric fever to those of typhus was so much larger, as to indicate that typhus had nearly disappeared from London ; and I have reason to believe that the same immunity from this form of fever has been observed in other localities where it was known to prevail epidemically. In the same years, enteric fever was the predominant disease. The average number of the last twelve years of this form was received into the hospital, the diminution in the total admissions having arisen from the comparative infrequency of typhus. If we trace back the records, we find that the number of typhus admissions, which in 1856 amounted to 1062, in the year 1858 had diminished to 15 ; and in 1860 did not exceed 25. During seven months of 1858, only a single case of typhus with the characteristic eruption on the skin was admitted. But towards the end of last year (1861) typhus again became epidemic, and from that time has spread rapidly, so that in January of the present year (1862) the number of admissions for typhus almost equalled that at any period of the history of the hospital, amounting to 140 ; and in the past six months the unprecedented number of 1098 has been received.

Of relapsing fever, essentially an epidemic disease, not a single case has been observed during the last seven years.

ing on this subject, allow me to make a few remarks on the history of enteric fever—the typhoid
affection of my distinguished friend M. Louis.

I have already drawn your attention to the
pathological differences between the one and the
other, and I am satisfied that enteric fever has
long existed in this country, though unrecognised
as a distinct form until a comparatively recent date.

While lately examining the pathological collec-
tion at the College of Surgeons, I discovered
several specimens of well-marked ulceration of
Peyer's patches, showing that the disease had not
escaped the notice of John Hunter. In the Pathological Museum of this College there are also
specimens of ulcerated Peyer's patches contributed
by Dr. William Hunter and Dr. Baillie, and which
are represented in the engravings illustrative of Dr.
Baillie's work on Morbid Anatomy, proving that this
form of fever, though it had not been described, was
not unknown in their day. But the morbid anatomy
of enteric fever was, as far as I have been able to
discover, first investigated by Rœderer and Wagler
in 1762; subsequently by Prost, a diligent patholo-
gist of Paris (in 1804); by Petit and Serres (in
1813); by Bretonneau of Tours, and his pupil M.
Trousseau (in 1826), and in 1827 by our country-
man Dr. Bright. In 1829 M. Louis published his
remarkable work,* so comprehensive and accurate
in details, as to leave little to add to the conclusions

---

* Edin. Med. and Surg. Journ., vol. liv.

he has formed; and although subsequent pathologists have taken no small credit for their researches,
they have only shown that enteric fever—this same
typhoid affection—is the endemic fever of the
United States, as it is known also to be that of
France, Germany, Switzerland, and a not infre-
quent form of fever in Great Britain and Ireland.

M. Louis, however, was evidently not acquainted
with our British typhus, as he admitted to Dr.
Christison and myself, when we had repeated interesting conversations with him in Paris in 1838.
He entered fully into the subject, and candidly
avowed his belief (which he re-stated in his second
edition) in the existence of another form of fever
—the British typhus, as he termed it—but which,
as he had never witnessed it, was unknown to him,
except by the description he had received of it in
his intercourse with British physicians.

Soon after my visit to Paris, Dr. Shattuck (a
favourite pupil of M. Louis, and now Professor of
Clinical Medicine in Boston, United States) came
to London, if not with the express object of in-
vestigating the pathology of fevers, this subject at
all events engaged a good deal of his attention.
At the request of M. Louis he visited the wards
of the Fever Hospital; and I well remember the
difficulty he felt in diagnosing the two forms
presented to his observation.

Among the more recent researches of many distinguished writers on the pathology of fevers, let
me draw your attention to a most valuable paper

by Dr. Stewart, physician to the Middlesex Hospital, which was read before the Medical Society
of Paris in 1840, with the object of showing the
distinct nature of the two forms of fever (typhus
and enteric).* It embodied results of investigations made in the Glasgow Fever Hospital, the
object being to prove, by comparison of the symptoms during life with the lesions after death, that
no two acute diseases can differ more widely from
each other, the characteristics of each being so
marked, if attentively observed, as to defy miscon-
ception. It certainly gave the first impulse in this
country to the investigation of the much-vexed
question—the identity or non-identity of typhus and
enteric fever; and I am satisfied that a perusal of it
will even now convey much practical information.

Again, from communications sent to England
by Mr. Scriven and Dr. Ewart, of the Bengal
Medical Service, both of whom having had the advantage of observing the two forms of fever in the
hospitals of London, are entitled to credit, we find
that enteric fever, as well as typhus, exists in
India and Burmah. Mr. Scriven has given the
history and morbid appearances of two well-marked
cases of enteric fever, in both of which the intestinal lesion was discovered.† Dr. Ewart gives the
history of six fatal cases which came under his
care at the Ajmeer gaol (Bengal)—four of enteric
fever and two of typhus—though from his observa-

---

* Edin. Med. and Surg. Journ., vol. liv.
† Medical Times and Gazette, vol. viii. New Series, p. 79.

tions it would appear that the rose spots so characteristic of the former are not generally visible
in the native Indians; not that they do not occur,
but they cannot always be seen through the dark-
coloured layers of the integuments. He states,
on this point, that there is no reason to believe
that the peculiar condition of the blood and capillaries which leads to the development of maculæ
is not invariably the same, whether the disease
(enteric fever) makes its appearance in a black or
a white subject. But the examinations made after
death put the question as to the nature of the two
forms of fever beyond dispute. Peyer's patches
were found more or less altered in the four cases
of enteric fever; in the other two (typhus) they
were normal.*

Now, in regard to my own views as to the forms
or types of fever—more especially with reference
to the question of the identity or non-identity of
typhus and enteric fever—I may state that the
work brought out by myself in 1828 (Clinical
Illustrations of Fever) was not in the first instance
intended for publication. It was suggested, in fact,
by my friend the late Professor Alison of Edinburgh, who, wishing to have more accurate details
than he had been able to obtain in regard to the
pathology of fevers as they were observed in London, requested me to furnish him with information

---

* Indian Annals of Medical Science, part vii., p. 65.

on certain points. Hence the origin of this bro-
chure. I admit that at that time, and for many
years afterwards, in common with almost every
physician in this country, I regarded those two
forms as identical, explaining the intestinal lesion
as one of the complications incidental to continued
fevers—in short, I believed that typhus and enteric
fever were not distinct diseases—and I continued
to entertain this view when I wrote the article
" Fever" in the Cyclopædia of Practical Medicine.

It is due to my own reputation, no less than to
the cause of truth, frankly to avow, that more
mature reflection, aided materially by investigations made in the wards of the Fever Hospital,
under my immediate inspection, by my friend
Dr. Jenner, and carefully weighing the arguments
and deductions of recent writers—British, Ame-
rican, and Continental—has led me to believe
that there are at least two distinct forms of continued fever, easily distinguished from each other
during life—the one by its invariably present specific lesions—the other by the equally invariable
absence of them. In short, that the two forms,
though having a close resemblance in the general
symptoms, and having many features in common,
are as distinct from each other as they are from
periodic or from exanthematous fevers.

Let me shortly lay before you the arguments
which appear to me to establish this proposition.

1. The symptoms and mode of access (as I shall

D

presently point out), though at first somewhat
similar, are more gradual in enteric than in typhus
fever. In enteric fever, the abdominal symptoms,
more especially diarrhœa, are almost invariably
present, while, after death, alteration in Peyer's
patches and corresponding mesenteric glands is
always found. In typhus, the invasion is more
sudden, the muscular prostration is more early and
more marked, the cerebral disturbance more constant and severe, while the gastric symptoms, so
predominant in enteric fever, are never observed.
The examination of the body, however, after death,
showing the absence of constantly present characteristic lesions, if there existed doubts during life,
is sufficient to decide the question. Not only are
the abdominal symptoms seldom if ever present,
but there is no instance on record on which the
slightest reliance can be placed, in which the intestinal lesion and corresponding mesenteric changes
have been found in fatal cases of typhus.

2. The duration of the symptoms is another
point of distinction, in respect of which the two
fevers differ materially. The duration of uncomplicated typhus may be assumed to be from fourteen to twenty-one days, that of enteric fever not
less than thirty days, though mild cases of either
may terminate at earlier periods. Again, typhus
runs a more rapid course, if uncomplicated, either
to convalescence or to a fatal result; it is, in fact,
considerably shorter, its progress more definite,

and it terminates more frequently by some evident crisis, such as sleep, or perspiration. Enteric fever, on the other hand, as a general rule, is of longer duration, and less regular in its course, some cases being short, others prolonged. Moreover, typhus, if it terminate unfavourably, generally does so in the second week; while enteric fever, though it may destroy life in the early stages, more usually carries off its victim from the third to the fifth week, or it may be even later.

3. The eruption peculiar to each form points strongly to their non-identity. The small, circular, bright-red, isolated spot peculiar to enteric fever, limited in most cases to the trunk, each spot being slightly raised, disappearing on pressure, and lasting for a few days, then disappearing, to be succeeded by a fresh crop or eruption,—forms a striking contrast to the more abundant, brick-dust, confluent, morbillous-like eruption, scarcely elevated, not fading on pressure, persisting until the period of convalescence, so characteristic of typhus. Indeed, the two eruptions are so different, so peculiarly attached each to its own disease, as to constitute a very important guide in the diagnosis—to the practised eye, quite as distinct as the eruption of measles is from that of scarlet fever.

4. Statistical investigations have shown a difference in the susceptibility to the two forms at certain periods of life. Enteric fever, we have seen, is, with few exceptions, limited to the middle

D 2

periods, being seldom observed (as a general rule) after the age of fifty. Typhus, on the other hand, is unlimited as to age, one-half of the cases being thirty years of age and upwards, and one-eighth above fifty. I have seen the disease in several persons above sixty, and it has been known to occur at the advanced age of seventy-five.

5. Arguments may be drawn, bearing closely on the question of the non-identity of the two forms, from a consideration of their supposed causes, which, even after the rigid scrutiny this department of the pathology of fever has undergone, are confessedly obscure.

Without attempting to discuss the etiology of fevers, I may observe that there is certainly an obvious difference in the origin of the two forms. For example, the possibility of typhus being engendered by over-crowding in infirmaries for the sick, in prisons, workhouses, transport ships, and dwellings of the poor, is a fact so well established, that no one in the present day attempts to dispute the statement. We have only to examine the records of epidemic visitations to be satisfied of the truth of this axiom; and, under whatever circumstances the disease may have arisen, its propagation by contagion explains its often rapid extension.

Again, famine and destitution are most powerful predisposing as well as exciting causes of typhus. I have already drawn your attention to various outbreaks of fever in Ireland during periods of scarcity and privation, in which it was often observed that the disease first attacked those who had suffered the greatest privations; and it may be conceived how rapidly it spread in hovels densely crowded with human beings under the most complicated misery. The form the fever assumed was invariably typhus, no cases bearing the least resemblance to enteric fever being noted in the reports. In short, united experience testifies to the fact, that epidemics of typhus have in general been preceded or accompanied by famine and destitution.

If we inquire into the circumstances that seem to favour the development of enteric fever, we find that the disease does not select the indigent alone for its victims, nor can any connexion be traced between it and destitution. We meet with it, on the contrary, in the habitations of the middle classes, and even in the palaces and mansions of the rich—generally in isolated cases—the subjects of it having been previously in good health, and in the enjoyment of every comfort. We must therefore look for other causes for its production; and, from the testimony of all who have inquired into this subject, it appears that whatever these may be, they are local and limited, not wide spread as in typhus.

It is asserted and believed by many whose opinions are entitled to consideration, that the emanations from decaying or decomposed organic

matter are the immediate cause of enteric fever;
and it would seem that whether the emanations
are the product of animal or vegetable matter, the
result is the same. Without attaching more weight
than they deserve to the well-known experiments
of Gaspard and Magendie, which have been often
quoted in proof of this point, I may simply state
the fact, that those physiologists were able, by injecting putrid fluids into the veins of dogs, to
induce symptoms in many respects resembling
those of enteric fever, and that on examination
after death, they found the mucous lining of the
small intestines in a state of congestion. These
experiments on animals do not, however, go for
much; they show merely that the introduction of
putrid matters into their blood is followed by
symptoms and morbid appearances supposed to
resemble, in some respects, those of enteric fever
in the human subject.

But on this subject—the invariable connexion of
enteric fever with emanations from decomposed
organic matters—I must confess my scepticism. I
am aware that on this point I differ from many
whose opinions are entitled to great consideration,
and I admit that there are some striking facts
tending to show that emanations from drains and
cesspools, when long confined, acquire, apparently,
concentration and poisonous properties capable of
inducing a powerful effect on those who become
exposed to their influence.

On the other hand, it must be remembered, that in some localities in which a few isolated cases of enteric fever have occurred, a very large proportion of the population have been exposed to the same supposed exciting causes—putrescent emanations—without apparently suffering in health; nay, the fever may cease entirely in such places without any sanitary measures having been adopted, and may not appear again even for years, though the same influences have continued in operation. Thus with the same pestiferous influences in full operation, there has been for the last few years a remarkable immunity from either form of fever, not only in and around this metropolis, but in other populous places (Edinburgh and Glasgow, for example) in which, in former years, one or other of the forms of fever has prevailed more or less extensively. Some other explanation of the origin of the fever-poison must therefore, in my opinion, be sought for.

I may also allude to the well-known immunity of nightmen and others employed in the emptying of cesspools. This necessary class of workmen almost invariably escape fever, and, indeed, seem often to fatten on their disgusting occupation. Mudlarks, too, who spend much of their time in sewers, seeking after articles which bring money, are known to resist this drain-poison in a remarkable manner.

But let me not be misunderstood in giving my

opinion on this important subject. I have always been a sanitary reformer, and supported the present laudable movement to improve our system of drainage, as well as the dwellings of the labouring classes; not that I conceive that the connexion between fever and defective drainage has been clearly established, but because I am satisfied that exposure to emanations from decayed organic matters, and more especially living in overcrowded dwellings, tend gradually to undermine the health by lowering the tone of the system, and that, when long continued, the principles of the blood become deteriorated. Under such circumstances but a spark is required to light up a fever, probably of a protracted and dangerous type.

It is an admitted fact, that enteric fever partakes more of an endemic than of an epidemic character, occurring in isolated places or dwellings, from causes often obscure and unexplained, and that while typhus does not appear to be materially influenced by the season of the year, enteric fever, though it may prevail at any period, is much more common in the autumn months. It seems also to be pretty well established that the two forms of fever —enteric and typhus—do not prevail together. Dr. Jenner traced to their respective localities all the cases of fever admitted into the Fever Hospital in 1847-48-49. In 1848, one-fourth of all the patients admitted had enteric fever; but out of forty-four localities, which furnished collectively

one hundred and one cases of typhus, there was but
one instance in which one patient with typhus and
another with enteric fever came from the same
house. Of the localities which furnished during
the same year nine cases of enteric fever, the same
statement held good. The same result happened
when the cases admitted in 1847 and 1849 were
traced. But the fact that one house had furnished
a case of typhus and one of enteric fever at the
same time, only shows that the general rule is not
without exceptions; so that if it turn out that the
special cause of typhus, whatever it may be, or
whencesoever derived, is peculiar to it—that is,
never gives rise to enteric fever—and conversely in
respect of the special causes of the latter (enteric
fever), a strong argument in favour of their non-
identity is established.

I may, in connexion with this subject (the
etiology of enteric fever), put a question rather
for consideration than for present discussion—may
not this fever have a spontaneous origin? I have
long thought that it may—an opinion that explains
the occurrence of many isolated cases under circumstances which render the supposition of their
origin in exposure to emanations more than
problematical.

6. Another important point of distinction is
the contagion of the two forms of fever. The propagation of typhus from person to person, from
room to room, from house to house, is so often ob-

served as to leave the question, as regards typhus, without a shadow of doubt. There are facts, on the other hand, tending to prove that enteric fever may be communicated from one person to another. The question, however, is more as to the degree of communicability of the two forms. It is no difficult matter to bring abundant proof of the contagiousness of typhus; but so rarely does enteric fever prove infectious, that unquestionable proof, on competent authority, is not often afforded. If we attempt to obtain conclusive evidence from hospitals, in which both forms are under treatment, the examples of enteric fever traceable to contagion rarely come under observation. I have certainly known a patient admitted into the wards with another acute disease—scarlet fever, or pneumonia, for example—contract, during convalescence, enteric fever; but it is a matter of observation that, while all who are on duty in the wards of the Fever Hospital are equally exposed to both forms, of the many I have seen attacked, the form has been in general typhus, very rarely enteric fever. Dr. Murchison has taken the trouble to ascertain from the records of the Fever Hospital the form the fever assumed among the nurses and medical attendants, and he found that in ten consecutive years, forty-eight cases of typhus originated among the inmates of the hospital, but only eight of enteric fever.

It thus appears conclusive, that the difference of the two forms, so far as their contagiousness as a distinctive characteristic is concerned, must have reference to degree; but certainly the highly contagious nature of typhus, when contrasted with the infrequency of the spread of enteric fever by contagion, ought, amongst other arguments, to have its due weight in the balance when the question of the identity or non-identity of the two diseases is debated.

7. The results of the treatment of the two forms should not be overlooked in discussing their identity or non-identity. As a general rule, we find that remedies have a much more striking influence on the one than on the other. Stimulants are more early required, and to a much greater amount, in typhus; and if any local complication arise in enteric fever, depleting remedies are much better borne than in typhus.

I remember Dr. Jenner remarking to me, as he investigated the records of the cases kept in the Fever Hospital, that when the leading characters of the two diseases had not been noted, or at all events so carefully as they have been within the last decennial period, he could, in the absence of the diagnostic symptoms, form a tolerably good idea of the type of the disease by the treatment that had been adopted; for while in the typhus cases, considerable quantities of wine and brandy had been prescribed, in the enteric, stimulants had

been either withheld or administered in sparing
amount.

Let me now allude shortly to the analogy between
these two forms and the exanthematous fevers.

I need scarcely recal to your recollection, that
under the exanthematous or eruptive fevers are
comprehended small-pox, measles, and scarlet
fever; and that the distinctive characters of this
group are—1, that they are contagious; 2, that, as
a general rule, they do not recur, that is, that
they affect individuals only once in their life;
and 3, that each is characterized by a peculiar
eruption.

Now let us see how far typhus and enteric fever
have, in all or any of their characteristics, an
analogy to the exanthemata.

My own opinion is, that they have a strong resemblance in many essential particulars.

As to the first point—their contagion—notwithstanding the doubts of a few, it may be assumed
as proved that typhus is a contagious disease, and
we have seen that enteric fever may, under certain
circumstances, be communicated. In this respect
consequently there is an analogy between the continued and the eruptive fevers.

The second point—that they are limited to a
single attack during life—is, perhaps, less easily
determined; but the evidence adduced in the
affirmative by trustworthy observers is such as to

render it more than probable, that those who have once had either form may be considered safe from another attack of the same fever. In some instances, individuals have been admitted into the hospital who had previously passed through an attack of enteric fever, and on close examination of the patient, and reference to the history of the first attack, recorded in the journals, it has been discovered, that the characters of the disease in both instances were identical; but such cases are so few in number as to form an exception to the general law, just as we know that during the prevalence of other epidemics (small-pox, measles, or scarlet fever, for example), the same person is occasionally attacked a second time.

The evidence of Chomel on this point is worthy of attention. He states that, notwithstanding the care with which patients affected with enteric fever were interrogated on this point on entering the hospital—and they amounted to 130—no one gave such a reply as would lead to the inference that there had been a previous attack; indeed, the almost invariable statement was, that it was the first time they had been ill; and he adds, that the result of the facts hitherto collected tends to prove that enteric fever, in ordinary circumstances, affects the same individual only once.*

Dr. Lombard of Geneva, in his observations on

---

* Chomel, vol. i. pp. 321, 333, and 309.

the difference between British and Continental
typhus, mentions that in one remarkable point of
view both agree—viz., that no one is known, or at
least is very rarely known, to have the eruptive
typhus twice. " With us (in Geneva) such instances are scarcely ever met with ; and I am in-
formed," he goes on to state, " that with you" (that
is, in this country) " a person once attacked with
typhus attended with the measles-like eruption,
may safely calculate upon immunity from the
disease for the future."*

The facts also adduced by Drs. Barker and
Cheyne give great strength to this view. At the
Cork-street Hospital, only one physician and the
apothecary had an attack of fever; but then, most
of the physicians had laboured under that disease
on some occasion previous to the epidemic. Similar
testimony is given by Dr. Davidson, in his Essay
on Fever. He tells us that almost all the clerks
and nurses of the Glasgow Fever Hospital, for
the last six or seven years, have had typhus
characterized by the eruption, " and not one of
them, so far as we have been able to learn, has
ever had it since, while almost all of them consider
themselves now perfectly secure against a second
attack, although constantly exposed to the effluvia
arising from fever patients."

The idea has been suggested by some physicians

---

* Dublin Journal, vol. x. p. 22.

—and it is a plausible one—that contagious typhus, like the exanthematous fevers, if accompanied by its characteristic eruption, produces in its course some changes in the system, by which an individual having once undergone the disease is, as a general rule, secured against a second attack. This theory has been argued by Dr. Penny of Glasgow, in his elaborate paper on Fever.* He thinks that in those instances of second attacks, and which form exceptions to the general rule, the fever appears in a mild and modified form, the crisis taking place on the seventh, ninth, or eleventh day. He has come to these conclusions, he adds, after careful observation of 4000 cases.

Others, again, Hildenbrand amongst the number, suggest that the miasm of typhus, after having produced the fever, destroys, almost always for a certain time, the susceptibility to a similar contagion, rarely, however, for the whole period of life, as small-pox and measles do.

The whole question is certainly full of interest in many points of view, and well worthy of further investigation.

The third point of analogy is the exanthema or rash. The distinctive characters of the rash peculiar to each form of fever have been already cursorily brought under your notice—the proportion of cases, as we shall afterwards see, in which it appears being so large, that the exceptions may be said to prove the rule. In the one—enteric fever—

* Edinburgh Medical Journal, vol. xiv. p. 87.

there is the lenticular rose spot, the period of the fever when it appears, its continuing a certain number of days, and then vanishing to be succeeded by a fresh crop to go through the same changes; while in the other—typhus—we observe the darker, morbillous, or mulberry eruption, differing in one important particular, that it is persistent so long as the fever lasts. Doubtless the occurrence of these two peculiar eruptions, dissimilar as in some respects they are, shows in both a marked analogy to the exanthematous fevers. The exceptional cases—those in which the eruption is absent—cannot be urged against this view, since in one of the eruptive fevers (the scarlet fever) the rash is occasionally absent, though in other respects the whole phenomena of the disease are manifest. Even in measles, the efflorescence is occasionally either absent, or so evanescent or ill-developed as often to lead to great difficulty and doubt in the diagnosis.

Besides these three points in which the analogy is well marked, I may allude to the duration of the two forms as constituting a point of resemblance. We know that in typhus, if uncomplicated, the symptoms decline about the fifteenth day, while the duration of enteric fever is much more protracted, being generally from twenty to thirty days, often much beyond. The eruptive fevers, when not complicated, run a pretty regular course, both as to the time of the appearance and decline of their distinctive rash, which we know differs in each.

It would seem, therefore, in their duration, as well
as in the several particulars thus briefly adverted
to, there is, to say the least, a great resemblance
between the two forms of continued, and the exanthematous or eruptive fevers.

We are now prepared to consider in detail the
various forms of fever according to the classifica-
tion exhibited in the tables.

I shall first commence with a description of
enteric fever, in order to enable me to point out more
clearly the circumstances in which it differs from
typhus.

ENTERIC FEVER.

Synonyms.—Anatomy of the intestinal glands.—Symptoms of enteric fever—of the mild form—of the more severe.—Symptoms considered in detail—in the nervous system—in the gastric organs.— Complications or secondary affections.—Latent enteric fever.— Duration of the disease.

ENTERIC fever has been described under different names—typhoid fever, typhoid affection, abdominal typhus, entero-mesenteric fever, gastric fever, ileotyphus. I prefer, however, the term enteric fever, on the whole the most simple, and expressing the prominent distinctive feature of the disease.

It is distinguished from all other acute diseases by its invariable anatomical character—specific lesion of the intestinal glands, and corresponding glands of the mesentery.

Let me in the first place draw your attention briefly to the system of intestinal glands.

Besides the follicles (crypts or corpuscles) of Lieberkühn, there are special glands in the intestines, which are usually described under three classes: 1, the glands of Brunner (or duodenal glands); 2, the solitary glands; 3, the aggregate or agminate glands (Peyer's glands).

The glands of Brunn, or Brunner, are arranged in the submucous or areolar tissue of the duodenum, and have the appearance of small granular bodies, to which Brunner gave the name of the second p ancreas, to the structure of which gland they bear some resemblance. They are most numerous in the vicinity of the pylorus, commencing abruptly at the duodenal side of the pyloric valve, and gradually diminishing in number towards its lower portion, until they finally disappear. These bodies, better distinguished by the name of duodenal glands, from their being limited to this portion of the digestive tube, are compound glands. They vary, both in number and size in different subjects, and seem to disappear altogether in advanced life.

The solitary glands have been sometimes, but erroneously, described as the glands of Brunner, from which, however, they differ materially in structure, position, and liability to disease. They are found in the form of small rounded bodies, of the size of millet-seeds, projecting upon the internal surface of the mucous membrane. They are met with in every portion of the intestinal tube, but in greatest numbers in the ileum, the vavulæ conniventes, and spaces between them; they occur also in the large intestine, particularly in the cæcum and appendix vermiformis. Around each gland there is a zone of the orifices of the follicles of Lieberkühn.

Examined by the microscope, they appear to be hollow sacs, covered with villi filled with mucus, resembling, on section, a florence-oil flask, wide at their blind extremity, tapering towards the surface of the mucous membrane, on which they open by a very small orifice, not easily seen, especially when the gland or sac is empty.

The agminate or aggregate glands are generally known by the name of glands of Peyer, who minutely described them, though they were previously known to and delineated by Dr. Grew, a Fellow of this College, in a paper published under the quaint title, "The Comparative Anatomy of the Stomach and Guts," being several lectures read before the Royal Society in the year 1676. Peyer did not publish his description till the following year (1677).* They ought therefore to be called Grew's glands, or patches, if priority of discovery regulate the question.

Each of these glands, the grouping or aggregation of which constitutes a single Peyer's patch, is in every respect similar in structure to that of the solitary, being a papilliform body, or vesicle, tapering to a pointed extremity, which projects amongst the tubes of Lieberkühn. They are arranged or collected together in oval or elliptical, sometimes circular, groups or patches, the long diameter corresponding with that of the intestine,

---

* De Glandulis Intestinorum.

and opposite to the attachment of the mesentery. The patches seem to belong, in an especial manner, to the ileum, and are found chiefly towards the lower portion, becoming more and more scattered, and less numerous in the direction of the duodenum, where, however, Peyer once found a single patch. The patches vary in number, from a few to twenty, thirty, or even more. Sometimes they assume the appearance of bands two or three inches long, or occasionally they form irregular clusters, the largest being near the ileo-cæcal valve.

It should be observed, however, that they vary much as to development in different subjects. In some, they are planted in the substance of the mucous membrane, to which they impart a degree of thickness; in others, they have the appearance of being embedded in the muscular coat. If the mucous membrane covering the patches be examined under the microscope, numerous distinct depressions may be discovered, each depression being the orifice of one of the follicles or crypts of which the patch is composed. It is not uncommon to find two or three diseased follicles in the middle of a patch which is otherwise sound.

Peyer's patches are most distinct in young subjects, and as they seem to undergo rapid change after death, it is necessary, if an accurate examination of their structure is to be made, that the bodies be recent, and, if possible, those selected in which the death has been sudden, or from accident, or from some form of acute disease that has proved rapidly fatal. It is generally supposed too, that the patches become changed, perhaps atrophied, from age, and after protracted illness, and it is not unlikely that the alterations they undergo with advancing age may have something to do with the infrequency of enteric fever after the age of fifty.

Of the use of the intestinal glands very little is known.

I have alluded to the resemblance in structure between the duodenal glands and the pancreas; and they are so strictly confined to the portion of intestine with which this gland is intimately connected, as to render it probable that the functions of both are of the same nature, and that they form part of the salivary apparatus of the intestinal canal.

The similarity in structure between the solitary glands and the glands of Peyer has been pointed out. Their functions, therefore, are probably not dissimilar. We know also, that the solitary glands and Peyer's patches both undergo important alteration in two diseases—tubercular phthisis and enteric fever—proceeding in both to ulceration, and to which the diarrhœa in the latter stage of the one, and, as a general rule, throughout the whole course of the disease in the other, is to be ascribed.*

The changes which these intestinal glands

* See Todd and Bowman, vol. ii. p. 234.

undergo, and which will be fully described hereafter, constitute the anatomical character—the
pathological element of enteric fever.

It may be naturally inquired how far this lesion
is to be regarded as the cause of the phenomena,
or in other words, is the febrile disturbance depen-
dant on, or symptomatic of, the local affection?
There can be little doubt, I apprehend, that the
intestinal affection is the result of the poison,
which appears to act simultaneously on the blood
and solids.

We observe in other acute affections—the exanthematous fevers for example—similar local
effects, induced by the action of a specific
poison. Thus, in small-pox there is pustulation
of the skin; in measles, affection of the skin and
mucous membrane of the air passages; in scarlet
fever, besides the cutaneous exanthem, the throat,
and in the more severe forms, the glandular
system, become implicated simultaneously. The
same compound action has been observed in the
plague, the poison acting probably first on the
blood, and subsequently on the absorbents, giving
rise to the well-known plague bubo. In the same
way the action of the poison which generates
enteric fever on the intestinal follicles is explained;
and whether it operates primarily on the blood or
on the nervous system, or on both, its specific
effect also on the intestinal follicles is scarcely
questionable.

After these anatomical and pathological details, we are prepared for a description of the phenomena observed in the progress of enteric fever.

It is unnecessary to state to my present audience, that though in many instances the symptoms in the early stage may be only slightly marked, and the progress of the fever mild, in others, and in by far the greater number, they assume a grave or severe character.

In the milder cases, the symptoms generally commence so slowly and insidiously, that the patient is often able to pursue his ordinary avocations, complaining only of undefined indisposition, such as irregular chills, loss of appetite, more or less headache, thirst, pain of limbs, and a degree of languor and lassitude, for which he cannot account. In the course of a day or two, these symptoms increase, the pulse quickens, there is less inclination for exertion, the tongue becomes furred, the bowels disposed to be relaxed, sometimes with abdominal pain; the skin irregularly warm, the face flushed, and at length, from increasing weakness, the patient seeks the couch, or prefers to keep to bed entirely.

Towards night, there is more or less aggravation of the symptoms, accompanied often with restlessness, inducing constant change of posture in the vain hope of finding relief; the sleep is disturbed and unrefreshing, the thirst more urgent, and the heat of skin more pungent. Sometimes, there is vertigo, or singing in the ears, occasionally, slight epistaxis.

Such are the symptoms of the early stage of enteric fever, and though to an experienced observer they are sufficient to indicate the disease, there may be reasonable doubt as to its nature until its characteristic eruption appears, which is seldom visible before the second week (generally from the eighth to the twelfth day) after the commencement of the symptoms. If the skin be closely examined about this period, a few small, circular, rose-coloured spots may be discovered, chiefly on the anterior and posterior aspect of the trunk, seldom on the face or extremities. As the characters of this specific eruption will be presently pointed out, I shall only remark that each spot is perfectly distinct, fades or entirely disappears on pressure, and after remaining visible for three or four days disappears, fresh spots coming out every two or three days, and undergoing a like process of eruption and decline. In this way, successive crops of these spots may be observed until the conclusion of the disease. This characteristic eruption, however, as we shall presently point out, is occasionally absent.

If the abdomen be examined, it will be found more or less distended and resonant, and a sensation of gurgling in the right iliac fossa, often accompanied with tenderness, is generally perceptible.

Diarrhœa is commonly observed as one of the

early symptoms.   It varies in degree, being sometimes
moderate, sometimes profuse and exhausting, and
in some cases accompanied with blood.

If the case be mild and uncomplicated, these
symptoms continue without marked variation till
towards the middle or end of the third week,
when a gradual abatement of the more prominent, especially of the diarrhœa, portends the
approach of convalescence.   The change, however,
is very gradual.   The pulse becomes slower, the
alvine discharges less frequent and more consistent,
the tongue more clean, the thirst abates, the heat
of skin and restlessness disappear, the sleep is
tranquil and refreshing, there is gradual improvement in strength, while the appetite for food
returns, and is often so keen as to require great
vigilance to prevent its too early indulgence, and
consequent interference with convalescence.

But let me remind you, that the progress of
enteric fever does not always run so smoothly.   On
the contrary, when the stage of convalescence is
daily watched for with great anxiety, all the
symptoms may continue unabated, or some of
them become even aggravated.   Thus the headache, which usually passes away towards the end
of the second week, may persist, and be eventually
followed by delirium, or by somnolence. gradually
lapsing into coma more or less profound; the
diarrhœa may increase, accompanied by a sudden

discharge of blood from the bowels, which, if it do not destroy life, adds much to the danger; or the intestinal ulceration may extend, inducing progressive emaciation and weakness. Again, the progress of the fever is not unfrequently interfered with by some form of pulmonary complication. This may be open and easily detected; but in some cases, more especially when there has been early and severe disturbance of the nervous system, it is latent, and discovered only by careful auscultation, which may reveal either bronchitis, pneumonia, or pleurisy, more or less diffused. Sometimes, and indeed not uncommonly in severe cases, in addition to the intestinal lesion, there may be both cerebral and pulmonary disease, so that the fever may be said in its progress to have involved the most important internal organs.

With the intensity of the local complication the febrile symptoms keep pace. The pulse rises in frequency, and becomes weak and compressible, the tongue dry, brown, shrivelled, and often fissured, the teeth and lips covered with brown or black incrustation, the emaciation progressive, the weakness day by day more marked; the evacuations passed unconsciously, the sacrum, hips, and other parts subjected to pressure, become inflamed, passing rapidly into gangrene; or pus may be deposited in different parts, more commonly in the joints, constituting pyæmia or purulent infection.

When the disease assumes a character so severe,

it is almost unnecessary to remark that there is little probability of a favourable termination; for though in some few instances, the patient may linger in an apparently hopeless state from day to day, and eventually recover, by far the greater number die from gradual exhaustion, about the end of the fourth week.

No doubt there are cases which, after struggling to a much later period, terminate favourably; indeed, there is scarcely a limit to the duration of enteric fever with severe local affection ; but when the symptoms continue beyond the thirtieth day, there is invariably progressive local disease to account for their persistence, so that it may be assumed, that local lesion sufficient even to be the immediate cause of death is always present when the febrile phenomena persist after that date, proving the natural duration of enteric fever to be about four weeks.

I shall now proceed to consider more in detail the individual symptoms of enteric fever.

Of the cerebral symptoms, one of the more early and constant is pain in the head.

It occurs in the mild as well as in the severe cases. Sometimes it is confined to the forehead and temples, more often it extends over the head, and is often accompanied by intolerance of light, conjunctival injection, or throbbing of the carotid and temporal arteries.

Louis affirms that pain in the head is one of the most common symptoms of acute fevers of every kind, but more frequent in persons affected with enteric fever than with any other disease; and as in the former (enteric fever) it is almost constant, and begins with the first accession of the symptoms, its absence at the beginning of a febrile affection, in which other symptoms characteristic of this disease are wanting, would indicate that the disease is not enteric fever.

Somnolence is another very common symptom. In the milder cases it is less marked, appears later, and is of short duration. In the more grave, it comes on early, is of longer continuance, and in fatal cases, it increases in intensity, until it ends in deep coma and death. Louis investigated this symptom with great minuteness, and found that out of fifty-seven patients (who recovered) eight had no drowsiness; in forty, in whom he noted with care the origin, duration, and degree of this symptom, in no individual did it occur on the first day of the disease, in only one on the second, in two on the sixth and eighth, generally on the ninth, and in an extreme case on the fortieth. Its average appearance was on the fourteenth day of the fever, its mean duration was eight days, and the extremes of the duration twenty-one days.

Somnolence is also one of the most marked symptoms of the enteric fever of children, and may be looked for at any period of the disease.

Delirium is more commonly present than absent in enteric fever. In a small proportion of cases it appears early—that is, within the first week, but in by far the larger number, much later—in the second or third week, or even so late as the thirtieth day. It is often transient, indicated only by confusion and incoherent answers, or when the patient awakes; in other instances, it is observed chiefly during an accession or paroxysm of fever, or in the night. In other instances, however, it comes on early and is persistent, seldom disappearing until the first signs of convalescence are apparent. In the still more grave, it lapses into coma, from which the patient rarely recovers; so that, as a general rule, early and persistent delirium, more especially if accompanied by drowsiness, portends great danger.

The character of the delirium varies. In some patients, there are sudden outbreaks of maniacal fury, in which there is great difficulty in restraining them. It is not uncommon in this acute form of delirium for the patient suddenly to leave bed, and to wander about the room or ward, or even to attempt to escape. In others, the delirium shews itself in fits of loud screaming, accompanied with incessant restless agitation, and a wild expression of alarm—a form of delirium of dangerous omen, especially if it appear early. In many instances, again, the whole phenomena have a close resemblance to delirium tremens; the patient mutters incoherently, is sleepless and agitated, the hands are tremulous or in constant motion, either picking the bed-clothes or catching at imaginary objects. This is often the character the delirium assumes when persons of irregular habits become the subjects of this fever.

The subject of delirium does not in general turn on any particular object, the incoherence being on unconnected points. Louis has alluded to a singular mental condition in some patients, who, while evidently suffering from the gravest symptoms, declare, on being questioned, that they are quite well—a kind of perversion of sensation and judgment which he thinks should always be regarded as one of the most alarming symptoms; indeed he affirms that he never witnessed a single recovery from it.

It has been remarked by all who have had much experience of the pathology of fevers, that however much the brain may have suffered, permanent mental imperfection rarely follows, and this applies to the form of fever we are now considering. Accordingly we find that although after severe cerebral disturbance, the powers of the mind are occasionally somewhat impaired, they are gradually restored as the patient regains strength, and this accords with the experience of Louis, who mentions that he had observed the intelligence affected during convalescence in one case only, but in this patient the mental capacity was small. He re-

mained for six weeks in a kind of idiotic state,
from which, however, he ultimately slowly recovered.

As to the relation between the delirium and the
condition of the brain after death, Louis tells us,
that, out of twelve subjects, who either had no delirium, or in whom this symptom occurred only mo-
mentarily during the last two or three days of life,
—in four, the cortical substance exhibited a slightly
rose tint throughout its whole circumference; in
six, it was perfectly healthy; in one case, it was
very much injected, with slight softening of one of
the optic thalami, the entire cerebral mass being
less consistent than usual.

So great a variety in the appearances after death
in patients who had no delirium, must lead us to
doubt our ability to find in appreciable alterations
of the brain an explanation of the symptoms of
which it is evidently the source, and to render
it more than probable, that those changes, when
they are found, take place in the last days of existence.

Although the cerebral symptoms in general seem
to bear no relation to, or are not dependent on, any
certain or fixed lesion of the brain or its membranes discoverable after death, I cannot concur
in the idea suggested by Louis, great as his authority is—that the brain symptoms are dependent
on the changes in the intestine, which constitute
the peculiar anatomical distinction of enteric fever;

and for this reason, that in many cases—I will not affirm in the majority—where the intestinal lesion has gone on and ultimately proved fatal, there has been neither delirium nor somnolency at any period of the disease.

I am disposed therefore to look, not to the solids, but to the blood, as the source of the disturbance of the functions of the brain in this fever, just as we find in uræmic poisoning marked sensorial affection. It may be said, that if the changes which the blood undergoes be the source of the cerebral symptoms, how are we to account for their absence in exceptional cases ? The explanation is, that there is a less amount of fever-poison in the blood—less than is sufficient to produce the brain affection; or it may be that, in some individuals, the cerebral mass is less susceptible, less easily impressed by the morbid blood,—just as we observe great difference in the effects of alcoholic stimulants on the nervous system in different persons.

In regard to spasms, this condition of the muscular system is not unfrequently observed in some form or other. There may be tremors, or subsultus, or starting of the tendons of the wrists (*carphology*), or twitchings of the muscles of the face, contractions of the diaphragm (inducing *hiccup*), and so on. In the latter stage of severe cases the spasms may affect the face, upper extremities, and diaphragm simultaneously.

F

Muscular debility, or prostration, is generally a prominent characteristic of both typhus and enteric fever, but it is more early and more pronounced in the former than in the latter—less marked, of course, in the milder cases. I have known many instances of enteric fever in which the strength was so little impaired, or the prostration so well concealed, that patients could not be persuaded that it was necessary to keep their bed, or even to be confined to the house, until they were compelled by the supervention of serious symptoms.

The symptoms referable to the senses require little more than enumeration.

In the eye or its appendages there does not seem to be anything peculiar, that is not observed in other acute diseases. Slight conjunctival injection is not uncommon.

The sense of hearing is often impaired, and it has been observed that if deafness come on at all, it is generally about the middle period of the fever. It usually progresses slowly; but even when it exists to an extreme degree, it does not materially influence the prognosis or final issue of the case.

Tinnitus aurium (buzzing or ringing) accompanied by deafness, but more frequently without it, is also an occasional symptom.

Towards the termination of enteric fever, inflammation of the external meatus sometimes takes place. It may end in resolution, but more often in abscess and purulent discharge. As it seldom spreads to the tympanum, loss of hearing is by no means a common result.

The sense of taste is generally altered or perverted, articles of food seeming to be either tasteless, or having a flavour different from what is usual. This may be due in some measure to the incrustation on the tongue and palate, but chiefly to the altered nervous sensation of the mouth and its appendages.

In some patients the sensibility is perverted. Augmented sensibility is not uncommon, more especially in females of the hysteric temperament, in whom the cutaneous and muscular systems being morbidly sensitive, the slightest touch causes apparently great pain. It is of importance to discriminate cutaneous tenderness in the epigastric and abdominal regions from peritoneal inflammation, for which it may be mistaken. The diagnosis is determined by observing that there is the same tenderness when pressure is made on any part of the body—the arm-chest, or lower limbs for example. On the other hand, we must not overlook the possibility of peritonitis springing up in the hysteric diathesis. In such mixed cases, the persistent vomiting, the thirst, hot dry skin, the attitude of the patient as she lies on her back with her knees drawn up, and the paroxysms of severe abdominal pain even when the belly is untouched, can leave little doubt as to the existence of peritoneal inflammation.

Muscular aching, back-ache, and pain or sense of

F 2

weariness in the extremities, are all referable to the same condition of the nervous system—perverted sensibility.

In the whole catalogue of symptoms diagnostic of enteric fever, none are so important, either in a distinctive or practical point of view, as those referable to the gastric organs. This may be anticipated from the lesions in this form of fever being chiefly limited to the abdominal organs.

In many patients, at the commencement of the disease, there are occasional feelings of nausea, which, if followed by retching or vomiting, may at first give an impression that the patient is only suffering from a bilious disorder. This idea is however rendered improbable by the little mitigation that follows the vomiting, as well as by the occurrence of other symptoms.

The sickness and vomiting of bilious fluid, generally accompanied by epigastric pain, may occur not only in the early stage, but when the fever is somewhat advanced. According to Louis, when these symptoms (epigastric pain and bilious vomiting) occur late, we may infer the existence of a lesion of the mucous membrane of the stomach. He found that of thirty fatal cases, twenty had nausea, vomiting, or epigastric pain, of whom eleven exhibited more or less serious alteration of the mucous membrane of the stomach, the extent of the lesion being, apparently, in proportion to the duration of the vomiting; while of fifty-seven patients, in whom the fever was more or less severe, but who ultimately recovered, forty-three had symptoms referable to the stomach—viz., thirty had pain in the epigastrium; nineteen, sickness only; and twenty had vomiting.

The tongue, in a considerable proportion of cases, exhibits little alteration. Louis found it natural, or nearly so, in about half the cases analyzed by him—that is, moist, without morbid redness, perhaps slightly coated with slimy whitish or grey fur—an appearance it may even preserve throughout the course of the fever In other cases, after having been slightly coated, it becomes more or less red at the edges and tip, and occasionally covered with a dark-brown stripe or fur on the dorsum; and as the symptoms progress, it is shrivelled, dry, and cracked or fissured, the fissures being transverse or longitudinal, and sometimes exhibiting ulcerations more or less deep. Occasionally the fur or coating becomes quite black, or a layer or stratum of blood is spread over the dorsum, while the teeth and lips are covered with a similar incrustation. This brown or black incrustation is due to exudation of blood, indicating an unusually severe form of fever, and more especially so when it is tremulous, and protruded and retracted with difficulty. There are, at the same time, dryness, heat, pricking, difficulty or pain in swallowing, and often irritating cough, due, in some degree, to the constant passage of the air over those surfaces, when the breathing through the nostrils has become impeded, in consequence of a congested or swollen state of the mucous membrane of the nasal cavity.

Though diarrhœa is not invariably present in enteric fever, it is so constantly observed as to constitute one of its characteristic symptoms. As in inflammation of the mucous lining of the intestines from ordinary causes, the follicular disease induces frequent action of the bowels, and hence the natural secretions, being evidently a source of irritation, are frequently expelled.

In mild cases, the bowels are little affected — seldom relaxed ; or if there be a tendency to re-laxation, it is generally in the advanced period of the disease. In the more severe, there is diarrhœa from the commencement and throughout the fever, the stools being serous or watery, often accompanied with griping, and followed, when the evacuations have been profuse, by great exhaustion. Sometimes they are ochre-coloured, or like pea-soup, often dark, resembling coffee-grounds, and occasionally mixed with mucus.

From Louis's statistics, we find that this symptom occurred in all his patients except three; and that out of forty cases observed, twenty-two had somewhat frequent and liquid dejections during the first day of the disease. Of the others, nine began to have diarrhœa between the third and ninth day and six between the eleventh and fourteenth. The three patients who had no diarrhœa died after the thirteenth and fourteenth days of the disease, and on dissection, ulceration of Peyer's patches was discovered. As to the number of evacuations, out of thirty-two patients in whom this symptom was accurately noted, it was found, that in eighteen, the average was eight to ten or more daily; in seven, the diarrhœa was more moderate, the average being four to six; and in an equal number—viz., seven— it was slight, the evacuations rarely exceeding two or three in twenty-four hours.

The diarrhœa, whether slight or severe, or whether or not it appears at the commencement of the disease, varies in its course. Sometimes, after gradually increasing, it becomes stationary; or it may diminish towards the latter stage of the fever. In other instances, such is its caprice, that after being moderate in the early stages, it suddenly increases in the advanced.

Again, the bowels, instead of being relaxed, may be confined, the evacuations being more or less consistent. In an excellent paper on Fever, by Dr. Wilks,* he says: "As a rule, the contents of the bowels in fever (typhoid—*i.e.*, enteric) are fluid, but not always so, as another case will show. It was that of a young girl, who had been ill three weeks, and whose bowels were said to have been irregular, but generally confined, and who came to Guy's Hospital to die. The post-mortem inspec-

---

* Guy's Hospital Reports, vol. i., New Series.

tion showed the intestine full of firm scybala; and on removing these, under each was found an ulcer." I have often met with similar cases, showing that there may be ulcerated intestine, and even perfora-tion of the bowel, without diarrhœa, instances of which will be found in the Clinical Illustrations of Fever published thirty years ago.

Intestinal hæmorrhage is peculiar, or nearly so, to enteric fever, for although it may occur in ordi-nary typhus, it is very rare. I had a case of typhus in the hospital some time ago, in which this appeared to be the destroying cause; and on open-ing the body, Peyer's patches were found to be sound, but the mucous membrane of the lower portion of the ileum, cæcum, and commence-ment of the colon, exhibited a red swollen appearance. In this patient there was the well-marked mulberry (typhus) eruption to assist the diagnosis.

The quantity as well as the appearance of the blood varies much; sometimes the amount is small, sufficient only to indicate the tendency to hæmorrhage, more generally, however, it is large, from perhaps half an ounce to several ounces; sometimes as much as twelve or even sixteen ounces have been discharged at once.

The evacuated blood differs in colour and con-sistence. It may be of a bright red, but this is not usual; more generally it is dark, or almost black, and as to consistence, it is generally thin and uncoagulated, resembling thick treacle ; occasionally, it is passed in solid coagulated masses.

The source of the blood may be traced to the ileum and cæcum, and to issue both from the ulcerated patches and intervening mucous membrane. In a case under my care in the hospital, examined after death by Dr. Jenner, it was discovered that water thrown into the superior mesenteric artery poured forth freely from the edges of an ulcer, from which, doubtless, the hæmorrhage, which was the immediate cause of death, proceeded.

Abdominal pain is rarely observed in the commencement of enteric fever, unless the abdomen, in the progress of the disease, becomes distended, when there is generally tenderness on pressure, especially in the right iliac region. It is often felt around the umbilicus, sometimes in the iliac region on either side, occasionally over the whole abdomen.

It is well to be cautious in examining the abdomen, especially in the later stages of this fever, in consequence of the activity with which the changes in the intestinal glands sometimes proceed, and the tendency of the ulcerative process to spread in depth, and ultimately perforate the peritoneal covering. The less, indeed, the bowel is disturbed the better, and hence the necessity for abstaining from aperients, and avoiding rough external pressure. I have already alluded to the hysterical tenderness not unfrequentiy observed in females of the nervous temperament, and the possibility of its being mistaken for peritonitis. In such cases the diagnostic hints I gave may be kept in view.

It is important to bear in mind the possible occurrence of a more serious form of abdominal pain — that arising from intestinal perforation. The characteristic signs of this lesion, which takes place in the advanced stage of the disease, or during convalescence, and often in cases apparently of no great severity, are, occasional chills or rigors, sudden excruciating pain in the abdomen, sickness and vomiting, marked change of features, clammy perspiration, rapid tympanitic distension, hurried breathing, and death within thirty to forty hours.

The maxim of Louis is full of practical import. He states that "if in the course of a severe or slight typhoid affection, or even under unexpected circumstances, the disease having been latent to that moment, there supervene suddenly, in a patient with diarrhœa, pain in the abdomen, aggravated by pressure, accompanied by altered expression in the features, and more or less speedily by nausea and vomiting, there must be a perforation of the small intestine." In this short sentence the symptomatology of intestinal perforation is succinctly given. Louis further affirms that he has never met with peritonitis in the progress of acute diseases, except as a consequence of perforation of the small intestine in subjects affected with the typhoid affection.

On the other hand, it should be kept in view that even when this accident (intestinal perforation) occurs, the symptoms may not be so well pronounced. For example, in a case given by Louis, in which, during cerebral disturbance, perforation occurred and was the cause of death, the only symptoms were, constant chills, and twelve hours before death a purple hue of the face, hands, and upper part of the chest. The abdomen, which was much distended, was not painful, except on firm pressure, which produced a little distortion of the features. On examination after death, a perforation about the size of a large pin's head was found in the ileum, about six inches from the cæcum.

It is well also to remember, that it may happen suddenly when it is little looked for. I have known it occur when convalescence was supposed to be progressing so surely and satisfactorily that the patient was allowed to leave the house. Nor is the state of the bowels, either as to the presence or absence of diarrhœa, to be depended on as a guide, for, as I have just pointed out, you will find cases recorded in which the evacuations were formed and healthy in appearance at the very time when the perforation happened, and destroyed life. Such instances, though fortunately not very frequent, show the necessity of giving a guarded opinion as to the ultimate issue of enteric fever, however apparently mild.

Though few patients survive more than from twenty-four to thirty-six hours (or a few hours longer) after the signs of perforation show themselves, instances have occurred in which existence has been prolonged for some days after. Louis had one patient who lived seven days after the development of the first symptoms, which were very violent at the commencement, and were not mitigated until the beginning of the fourth day. In a case under my care in the hospital, the patient survived sixteen days.

Meteorism or tympanitis is a common symptom in enteric fever, and seems to arise from some special cause, as the condition of the mucous membrane of the colon, in which the air is chiefly collected, does not reveal its origin. The amount of accumulated air is sometimes moderate, sometimes, however, excessive. In mild cases, it is either absent, or so slight in degree as not to produce sensible elevation of the abdominal parietes, and may not even be perceptible on palpation. In the more severe, it is always present, the degree or amount corresponding with the gravity of the other symptoms; and when the distension is considerable, the discomfort is much increased by the impediment offered to the descent of the diaphragm. In a recent instance under my care, for example—and I have witnessed several such—the tympanitic resonance was well-marked as high as the third rib, and produced the greatest distress. I succeeded in giving relief and freedom to the lungs and heart, by having an œsophagus tube introduced into the colon, through which the air freely escaped. The tube was kept in this situation for many days, being only removed from time to time. As I shall afterwards speak of the mode of managing this symptom, the present allusion is made merely with the view of pointing out the enormous distension and corresponding distress occasionally induced by the accumulated air.

As to the period of the fever when meteorism supervenes, except in severe cases it seldom appears early, generally not until the second week, sometimes much later. In one instance given by Louis, it was not observed until after thirty-eight days; and in another, not until the fever had run on so long as sixty-five. When it is present, it continues until the termination of the disease.

In regard to its comparative frequency, Louis noted it in thirty-four out of forty-six cases which were fatal, and in forty out of fifty-seven patients who recovered; it began in four on the seventh or eighth day; in one, on the ninth; in thirteen, between the tenth and twelfth; in a majority, at a much later period; while in some, in whom the fever ran its course slowly, it did not occur before the third week, or even later. As to degree, it was considerable in seven, and moderate, or very slight, in the remaining; it diminished gradually in many, rapidly in others.

Though it may appear unaccountable that this symptom—meteorism—should be mistaken, such a diagnostic error has been committed. For example, Louis states that he heard an eminent physician pronounce a patient with the typhoid affection (enteric fever) to have hepatization of the right lung. In this case, the abdomen being greatly distended with air, a flat sound was elicited, much higher than usual, on percussing the right back of the chest. Examination of the body after death showed the lung to be quite healthy, and the cause of the flat sound to be the liver, which had been pushed very high by the great distension of the tympanitic colon.

During my attendance, some years ago, on a gentleman with enteric fever, an hospital surgeon was summoned in consultation, in consequence of the alarm excited by the unusual tympanitic distension. This gentleman mistook the hour appointed for our meeting, and after waiting some time, the medical attendant of the family, who had watched the case with me, was sent for, and the examination of the patient was proceeded with. On retiring to an adjoining room, the pathology of the case was debated, when this surgeon came to the conclusion, that the disease was ascites, and, notwithstanding the remonstrance of the family attendant, was in the act of prescribing elaterium, when I arrived, and took the liberty of stoutly resisting both his diagnosis and therapeutics. The patient after a protracted illness ultimately recovered.

In the enteric fever of children, this symptom occurs not less frequently than in adults. M. Rilliet met with it, in various degrees, in about two-thirds of his cases, but more frequently in those who died, than in those who recovered.

On the whole, then, it may be inferred, that the infrequency of meteorism in acute diseases, not enteric fever, and its frequency and degree in the latter, render it an important diagnostic symptom.

If the region of the cæcum be carefully examined by the fingers of both hands—with the precaution, for obvious reasons, of handling this portion of the intestinal canal with care—a gurgling sound may be elicited. It was first pointed out by Chomel, who considered it as one of the signs of enteric fever. Though it may be frequently detected, it is much oftener absent, and cannot therefore be regarded as one of the constantly present diagnostic symptoms of this disease.

Let me now draw your attention to one of the most distinctive characters of enteric fever—not, however, as we shall see, invariably present—an eruption of small lenticular rose-coloured spots,

scattered chiefly over the surface of the abdomen
and chest, where the ·cuticle is known to be
thinnest. Sometimes a few of these spots (from
six to twenty) are seen, after careful examination,
on the abdomen, or one or two may be discovered
on the chest and back; though when there is an
abundant eruption, they are not limited to the
region of the trunk, but may be seen on the extremities, and even, though very rarely, on the
face.*

Each spot is of a circular shape, varying in size
from a point to a line and a half, rarely exceeding two lines, in diameter, slightly raised above
the surrounding cuticle, and of a pink-rose colour,
disappearing on pressure, but returning as soon as
the pressure is removed. Their eruption is not
attended by any unusual sensation, and they
seldom appear before the fifth day of the fever,
more commonly in the second week.

If, however, an average be taken, these rose-
spots may be assumed to come out between the
sixth and ninth days after the commencement of
the fever; it is certain, at least, that in half the
number of patients they are found between those
days. It would also appear that, except in cases

---

* For permission to exhibit a beautiful wax model of this eruption, executed by Mr. Town, whose artistic skill has enriched the
museum of Guy's with the most valuable series of pathological models
in Europe, I was indebted to the authorities of Guy's Hospital.

I have succeeded in obtaining a drawing of this eruption, which
is a faithful representation of nature.

of relapse, fresh rose-spots do not appear after the thirtieth day of the fever.

Again, these spots do not, as in the typhus eruption, persist throughout the fever, but, after remaining visible eight or nine days, they gradually fade, and ultimately entirely disappear, without leaving a mark or stain. They are, however, replaced by a fresh eruption, which, after pursuing a precisely similar course, in its turn dies away; and in this manner the spots come out and disappear, until the termination of the fever. Their mean duration may be stated to be from eight to nine days.

M. Louis, who observed these spots very carefully, found them readily in twenty-six out of forty-six cases that were fatal; and in fifty-seven that recovered, they were discovered in all except three. They were also noticed to be invariably present in all the cases that were mild.

In respect of the period of the disease at which the eruption came out, in two patients, it was visible on the sixth day; in three, on the seventh; in a third of the cases on the tenth; and in ten, it did not appear until between the twentieth and thirtieth day. In the cases tabulated by Chomel, it was noted in two between the sixth and eighth day; in thirteen, between the eighth and fifteenth; in seven, between the fifteenth and twentieth; in four, between the twentieth and thirtieth; and in one, on the thirty-seventh day.

G

This eruption (of rose spots) is, according to
M. Taupin, not less frequently observed in children
than in adults; for out of 121 cases of enteric
fever in early life noticed by him, it was present
in all except eleven, and of these eleven, three were
not received into the hospital until three weeks or
more after the commencement of the disease, during
which period the spots might have disappeared.
M. Rilliet, on the other hand, observed them in
two-thirds of his cases only; but he does not say
whether the patients, in whom the spots were not
seen, were received into the hospital at a late period
of the fever. It may, however, be assumed that in
children, the spots appear at an earlier stage of the
fever than in adults, for both these writers state,
that they observed the eruption generally from the
fourth to the eighth day, sometimes sooner, occasionally later. They were found by M. Taupin in
some children as early as the second day, but they
generally disappeared after a day or two.

It is satisfactory to find from this author, that
he had never met with the rose spots in any other
infantile disease, except enteric fever.

There is another point in connexion with this
spotted eruption that requires passing notice. In
some cases it is preceded for a day or two by a
scarlet efflorescence, very much like roseola. When
this roseolous rash appears, it is apt to throw doubt
on the nature of the disease, and there is even a
possibility of its being mistaken for scarlatina. As

the disease progresses, however, the doubt is soon dispelled.

But I have stated that this characteristic rose-coloured rash is not invariably present. It is very difficult to obtain statistical results on this point that can be relied on. The spots may have disappeared, when the patient is examined, or they may be so few in number as to escape detection. This point has not been overlooked in the Fever Hospital, for the presence or absence of the characteristic rash is invariably noted; and were I to give an approximative idea of the total number of cases in which the rose spots do not appear at any period of the fever, I think ten or twelve per cent. will be nearly correct.

In the progress of enteric fever, we occasionally observe, too, a vesicular eruption, termed, from its semblance to sweat drops, sudamina, though no one can suppose that it has the most remote connexion with the function of perspiration. It is known also by the name of miliaria, or miliary rash, and consists, as you are aware, of small, prominent, round, transparent vesicles, formed by the effusion of limpid fluid under the cuticle. The vesicles are usually about the size of millet-seeds, the intervening spaces retaining the natural colour of the healthy skin.

This rash, which is an occasional accompaniment of other acute maladies, rarely appears before the end of the second week of the fever. It comes out more generally on the anterior aspect of the thorax, on the abdomen, inguinal regions, neck, and axillæ; seldom on the back or limbs, and never on the face. Its duration is not uniform; sometimes the vesicles, after continuing prominent for a few days, shrivel, and disappear with desquamation; in other cases, they remain visible for eight or ten days. Their average duration, however, may be stated to be from three to ten days.

This eruption does not seem to have any special influence on the symptoms or progress of the fever, though it has more frequently been observed in cases of severity than in the milder. And as to its comparative prevalence, out of ninety-eight patients (observed at the Hôtel Dieu or La Pitié), it was discovered in seventy-six; and it was remarked, that there seemed, in these instances, to be very little relation between the abundance of the sudamina and the condition of the skin as to perspiration.

This eruption is also observed in the enteric fever of children, M. Rilliet having noted it in two thirds of his cases, and M. Taupin in one hundred and four out of one hundred and twenty-one. M. Rilliet states that it generally appeared between the eleventh and twentieth days, and that its average duration was from one to six days.

I may here advert to the occasional appearance of erysipelas in the progress of enteric fever, which, however, is to be regarded as accidental or intercurrent only. It shows itself in the wards of the Fever Hospital at certain periods of the year, just as it is known to prevail in other hospitals amongst the medical, but more especially amongst the surgical patients; often appearing suddenly, lasting for a time, and then disappearing. It has an apparent connexion with trivial wounds, such as leech-bites, cupping scarifications, or the application of a blister; at other times it appears as a constitutional affection, commencing, perhaps, in the throat, and spreading outwards, through the nose or ear, to the face, scalp, and neck. It seems to have no precise or definite duration, and terminates either in resolution, vesication, or purulent infiltration, and, when severe and extensive, may be, *per se*, the destroying cause. Accumulated facts tend moreover to shew that it may be propagated by contagion.

Epistaxis is another occasional accompaniment of enteric fever. The amount of blood varies, being sometimes small, perhaps only a few drops; more often it is moderate, seldom profuse. It may occur once only, or for several successive days, nor is it limited to any period of the fever, occurring sometimes at the commencement, but more commonly when it is somewhat advanced. It is seldom preceded by sensations denoting its advent, nor is the patient sensible of any change in the symptoms after its cessation.

It is observed less frequently in the mild than

in the more severe cases. Thus, Louis noted it in about half the cases that were mild, and in twenty-seven out of thirty-four in whom the fever was severe, but not fatal. Out of the cases observed in the Fever Hospital, the particulars of which were noted from the commencement of the disease, it occurred in one-third only.

As to its occurrence in childhood, it happens oftener before than after fifteen years of age. M. Taupin, for example, met with it only three times amongst one hundred and twenty-one children, and in those only to the extent of a few drops of blood; and he noticed, that several children who had been subject to bleeding at the nose before they were attacked with fever, had none at all during the whole period of their illness.

I have thus given in detail the symptoms by which enteric fever may be recognised; indeed they may be said to be pathognomonic of the disease. There are, however, other organs and parts of the body which occasionally become involved in its progress; and it has been usual to describe these as complications, or superadded affections. We know, for example, that although in this fever the pulmonary organs are not necessarily involved in the general disturbance, there are few cases of severity in which they escape.

The pulmonary complication may be exhibited either in the form of simple catarrh, bronchial

catarrh (bronchitis), inflammation of the pulmonary parenchyma, or of the pleural covering, each
being indicated by its peculiar local or physical
signs.

In a considerable number of patients, but especially when there is coexisting brain affection, the
pulmonary lesion is latent, and can only be detected
by careful auscultation. Irregular accessions of
flushing and heat of skin, especially if accompanied
with paroxysms of coughing, are sufficient to indicate some form of secondary pulmonary affection, the precise nature of which can only be revealed by physical examination. But such general
symptoms are often absent, and the lung disease
may remain undiscovered, unless the practitioner
is on the watch, and, by repeated auscultation,
detects its existence; and it should be kept constantly in mind that the pulmonary lesion may
suddenly spring up at any period of the fever,
and render a mild case severe, or even dangerous,
if not fatal.

The bronchial catarrh, however, is perhaps the
most common form of pulmonary complication in
this country, and is indicated by frequent cough,
glairy tough expectoration, wheezing respiration,
and moist râles over the chest.

Pneumonia is a much more serious affection.
Rokitansky seems to think that in all cases of
severe enteric fever, but especially when the
symptoms are well marked, there is hypostasis in

the inferior portion of the pulmonary tissue, which
passes into a low form of pneumonia and subsequent
effusion of a gelatinous soft product, similar to that
thrown out on the bronchial and intestinal surfaces,
and corresponding with the existing typhus dyscrasia, which he supposes to be the result of an
adynamic state of the system.

This pulmonary lesion is indicated chiefly by
the local signs—the dulness on percussion—the
fine crepitation at first, and afterwards the bronchial breathing, the consequence of portions of the
lung being blocked up by the infiltrated morbid
products.

Angina laryngea is fortunately not a frequent
secondary affection in enteric fever. I say fortunately, because it is generally fatal when well
marked. It arises from sudden and rapid inflammation of the glottis, which terminates in
œdematous infiltration in this delicate structure.
Rokitansky ascribes it to the so-called typhous
deposit in the larynx. It may be recognised by its
characteristic symptoms — intense agony in the
effort of swallowing, succeeded by ringing cough,
stridulous sound of the voice and respiration, and
frequent attacks of laryngeal spasm, threatening
suffocation. I have never witnessed more frightful
suffering than this symptomatic angina, when fully
established, exhibits; while the look of despair in
the patient's countenance abundantly testifies to
the reality of the agony endured. If not arrested

in the very early stage, it runs its course to a
rapidly fatal termination.

In enteric fever, there is a marked tendency to
inflammation of serous membranes, and hence
pleuritic inflammation is a frequent complication
or secondary affection.   It may not always be
recognised by the general symptoms alone, but by
careful auscultation its presence can scarcely be
overlooked.   It would appear to be less frequently
observed in the typhoid affection of Paris, for
Louis mentions that out of fifty-seven cases, he
only met with one example of it.   This infrequency
of pleurisy as a complication of enteric fever does
not accord with the experience of British physicians; and from the susceptibility in this fever to
serous inflammation, more especially to pleurisy,
it is well to be on the look out for it even in mild
cases.   It is indicated by sudden chills or shivering,
pain or stitch in the side on deep inspiration, disinclination to lie on the affected side, hurried
breathing, pleural friction sound, with circumscribed dulness.   These characteristic signs, however, are not always present; indeed, in the greater
number of cases, the disease is less open; there may
be no rigors, little if any pain or stitch, little or no
hurry in the breathing, perhaps only flushing and
hot skin, to indicate that something is wrong.   Examination of the chest seldom fails to reveal the
secret, by creaking heard over the seat of the pleuritic inflammation; or if the mischief has crept on

still more insidiously, the percussion note is dull and
the respiratory murmur absent, indicating the existence or fluid more or less copious in the pleural
cavity. It is remarkable how rapidly the fluid accumulates in some cases, and even with little warning
of the antecedent inflammation; hence the necessity
of watching, in cases of enteric fever even comparatively mild, for the first approach of pulmonary
symptoms, which may surprise us at any stage of
the disease, or even during convalescence.

When the pleuritic inflammation occurs on the
left side, it may extend to the pericardium and
even to the valvular apparatus of the heart—a
complication which, I need scarcely observe,
though not necessarily fatal, seriously involves the
safety of the patient.

Again, it should be kept in view that fever
(either enteric or typhus) occasionally attacks
persons who are the subject of chronic pulmonary
diseases, just as those suffering from other organic
affections may become its victims. But if the
history of each case be carefully inquired into, it
will seldom fail to distinguish this class from those
of enteric fever with acute pulmonary complication,
though certainly, in some cases, it is difficult to
arrive at a correct diagnosis.

In the abdomen, peritoneal inflammation as a
secondary affection is occasionally observed, but is
certainly not common. When it comes on suddenly,
and with great intensity, the possibility of its

originating from intestinal perforation should be kept in view.

The other lesions found in this cavity after death —the various changes in Peyer's patches, in the mesenteric glands, and in the spleen—to be afterwards more particularly described—are to be viewed, not as complications, but as constituting the anatomical elements of this form of fever.

The various secondary lesions or complications which occur in the progress of enteric fever, not only influence materially the general progress and character of the disease, but may be even the immediate cause of death. The lesions in question are to be ascribed to the specific effect of the fever poison on individual organs, in the same manner as other poisons intoduced into the system affect particular parts.

In regard to the brain affections, it is important to bear in mind that symptoms precisely analogous may arise from very opposite pathological conditions—headache, delirium, somnolence, and convulsions supervening, and even proving fatal, without abnormal vascularity being discovered after death; and although cases do occur in which there is evidence, on examination, of pre-existing vascularity, indicated by redness of the membranes or substance of the brain, or of both, or by effusion of serosity or of lymph, such instances are much less frequent than those in which the appearances after death by no means correspond with, or account for, the local disturbance during life.

The pulmonary complications, too, though exhibiting the ordinary physical signs by which the idiopathic forms are recognised, partake less of an acute character; and if we adopt the opinion of Rokitansky, they are mainly due to the infiltration of that low or caco-plastic material he has denominated the typhous process, by which a temporary congestive action is set up, accompanied by increased constitutional disturbance.

In their treatment also, as I shall afterwards point out, due regard must be had to the peculiar nature of the local changes, and the condition of the system in which they arise.

If we go to the great centre of the circulating system—the heart—so constantly and variously disturbed in fever, we seldom if ever find the slightest evidence of inflammatory action; on the contrary, its action may be so weakened that it appears scarcely equal to its ordinary work, and hence its sounds, especially the first, are nearly inaudible. It is in this condition of the heart that we often find the patient rambling or unconscious; the face and lips pale, and the surface cool, or perhaps cold and livid. And this bloodless state of the brain in such cases explains the quantity of alcoholic stimulants occasionally required, both in enteric and in typhus fever, to rouse the dormant powers, and the tolerance of quantities which, if taken by the same individual in his ordinary or healthy state, would deeply intoxicate.

As to the condition of the abdominal organs, if we except the specific lesion of the intestinal glands, mesentery, and spleen, we rarely find disease either of the peritoneum or other structures within this cavity. It is true that, when the typhous process has destroyed in succession the coats of the intestine, and the perforating ulcer permits the alimentary contents to become effused into the abdominal cavity, violent peritonitis and rapid death follow. But this is an accidental lesion only.

Let me for a moment draw attention to latent enteric fever. We have been hitherto considering the more ordinary course of this disease, the symptoms of which are generally, in prominent particulars, so well marked as to be readily recognised—at all events by those who are practically acquainted with the pathology of fevers. Cases, however, occasionally occur, in which, though some of the symptoms may be present, the general aspect of the disease is such as to leave its nature somewhat uncertain. For example, medical aid is sometimes requested by an individual, who is unable to give a better description of his ailments, than that for some time previously he has been suffering from undefined indisposition, which he is unable to throw off; he complains of lassitude, irregular chills alternating with transient flushes of heat, of being easily fatigued, and unable to pursue his usual employment, and of loss of appetite and restless nights. On further inquiry, it will transpire that the bowels have been irregular, with tendency to griping and purging, and consequently the whole ailment has been ascribed, not unreasonably, to what is popularly known as bilious disorder. After a few days, it is manifest that matters do not improve, but are probably getting worse; the prostration and disinclination to exertion have so much increased, that there is no longer the desire, or even the ability, to keep about, and the patient of his own accord remains at rest on the sofa or in bed. Some, more energetic and enduring than others, make an effort to pursue their usual avocations, and I have known individuals transacting in a measure the ordinary duties of life,—mercantile men appearing "on 'Change," or clergymen performing their parochial duties,— when on close examination, besides the altered countenance, general excitement, and staggering gait, the peculiar rose-spots, relaxation of the bowels, and abdominal distension, showed too clearly the serious nature of the latent or hidden disease. I have known instances in which the patient could not be persuaded that he was seriously ill, until profuse hæmorrhage from the bowels convinced him, that the apprehensions of his friends and medical attendant were not groundless; and I have seen, moreover, such cases terminate fatally a few days after the first alarm was taken.

Again, the disease may assume the form of common cold; there may be cough, accompanied by moist râles over one or both lungs, in addition to other symptoms; so that the whole phenomena are ascribed to the patient having " taken cold." But in a few days, the chest affection passes away, leaving the other symptoms stationary, or, it may be, somewhat aggravated, until the true nature of the disease becomes revealed by some of the prominent signs of enteric fever.

I need scarcely remark to my present audience that such latency is not confined to the class of maladies now under consideration, but is observed in other acute affections, more especially subacute pulmonary inflammations—pleurisy and pneumonia for example—in which the only general signs to indicate the one or the other may be, perhaps, occasional cough, slight acceleration of breathing on unusual exertion, and irregular paroxysms of fever. When the chest is examined, we find one side dull, perhaps as high as the scapula, or even the clavicle, with absence of respiratory murmur; and, if the effusion be considerable, prominence of the intercostal spaces, if not perceptible dilatation of one side.

But in uncomplicated latent enteric fever, physical diagnosis does little towards clearing up the mystery, so that we are obliged to trust to a minute survey of the symptoms, and to contrast them with whatever previous history we can obtain; and, after all, we are too often in the dark, and obliged either to confess our doubts and fears, or make a bold guess as to the nature of the disease. Happily, however, such obscurity is comparatively rare, and I have drawn your attention to these cases of latent enteric fever more with the view of warning you of their occasional occurrence, than of aiding you with the diagnostic signs.

M. Louis, in alluding to the occasional latency of this form of fever, gives several marked cases in illustration, and has shown that the disease may proceed even to intestinal perforation, without any of the symptoms that usually indicate this fatal lesion.

Another point in the pathology of enteric fever requires notice—its duration. The average duration of this fever is from twenty to thirty days, that is, if no local affection or complication supervene; and it appears from united experience, that if it exceed thirty days, some local cause exists to account for the protracted duration.

M. Louis, after analyzing the cases observed by him, which proved fatal after the thirtieth day—nine in number—found in all of them local lesions sufficient to account for the fatal issue, which accords with the results obtained at the Fever Hospital; not a single case fatal after the thirtieth day having been noted by Dr. Jenner, in which, after inspection of the body, there was not a lesion discovered sufficient of itself to destroy life.

PATHOLOGICAL ANATOMY OF ENTERIC FEVER.

*Lesions in the pharynx—œsophagus—stomach—duodenum—ileum
—colon—mesenteric glands—mesocolic glands—spleen—liver—
kidneys—salivary glands—submaxillary and sublingual glands
—brain—medulla spinalis.*

HAVING considered the symptoms which charac-
terize enteric fever, I shall next bring under review the pathological anatomy or anatomical characters of the disease.

The true pathological element of the disease
consisting of specific lesions in the alimentary
canal and corresponding glands of the mesentery,
it seems the more proper course to point out, in
the first place, the changes that take place in the
abdominal organs.

In the pharynx the mucous membrane occasionally exhibits the ordinary results of congestion,
or it may be of inflammation, in redness more or
less deep and swelling more or less diffused, with
patches of fibrin here and there on the posterior
pharynx, velum, uvula, and epiglottis. Sometimes
an ulcer, superficial or deep, is found in the posterior wall of the pharynx; or there may be two
or three smaller ulcers around one larger. In other
cases we find an ulcer on one or both tonsils, or the
mucous membrane of the epiglottis destroyed, exposing a superficial ulceration, with effusion of
muco-purulent fluid in the sacculi of the larynx.

The only alteration met with in the œsophagus

H

is ulceration, and this is not frequent. The ulcers
are small—from two to twelve lines in the longest
diameter, oval-shaped, generally superficial, rarely
deep. When few in number, they are found near
the cardiac, or towards the middle portion of the
œsophagus; if they are numerous, they are scattered over its whole length, but in greater numbers
near the cardiac end, in which situation they are
also of larger size. It has been observed, too, that
in the majority of cases in which these ulcers are
found there is also some lesion in the stomach.

In one case that occurred in the Fever Hospital,
reported by Dr. Jenner, the œsophagus was extensively ulcerated. The ulcerations, several in
number, extended five and a half inches upwards
from just above the cardiac orifice of the stomach.
The circular muscular fibres were at places exposed. One of the ulcers was three inches in
length, and half an inch in breadth. All were
longer than they were broad. The edges of the
majority were slightly elevated. The intervening
mucous membrane was here and there finely injected. This was the only case, out of seventeen
examined, in which the œsophagus was diseased.

In rather more than half the cases examined
after death, the stomach is found to have undergone some alteration, chiefly, if not exclusively, in
the mucous lining. The most common alterations,
according to Dr. Bartlett,* may be arranged under

---

* History, Diagnosis, and Treatment of the Fevers of the
United States.—By ELISHA BARTLETT, M.D.

changes in colour, consistence, thickness, mamellonation, and ulceration, which may exist separately,
or, as happens more frequently, two or more of
them are found together.

As to colour, the natural pale colour of the
membrane is sometimes changed into a deep pink
or even red tint, most apparent towards its great
curvature; but whether the various degrees or
shades of redness be connected with previous inflammatory action has not been determined.

Softening of the mucous membrane, either alone
or combined with diminished thickness, is occasionally observed near the cardiac extremity, the
mucous membrane being found in some cases entirely destroyed, and the other coats thin and
lacerable. Chomel found softening near the cardiac extremity in ten out of fourteen cases. Louis
observed it (with diminished thickness) in nine
out of the forty-six cases examined by him,
and states, that it may be expected in all cases
which are fatal before the twenty-fifth day of the
fever.

Mamellonation, as described by Dr. Bartlett, consists of small elevations of the mucous membrane,
pretty regularly circular or oval, scattered thickly
over different portions of the stomach, but more
commonly near the pylorus. This peculiar condition generally exists in connexion with other
alterations, especially with softening and increased
redness. Louis found it in thirteen, or two-

sevenths of his cases, occupying nearly the whole
surface of the stomach; and, as it was much more
frequent amongst those who died between the
eighth and the twentieth days of the fever, than at
a later period, it may, like all the morbid changes
of this membrane, be considered to be more frequent and more severe according to the more or
less rapid course of the fever.

Ulceration of the mucous membrane of the
stomach, distinguished from destruction of the
membrane from softening, by the small size, exact
limitation, and sharp, clean edge of the ulcers, is
very rarely met with in enteric fever. When
observed, these ulcers are superficial and vary in
number, but as many as twenty have sometimes
been noticed. To show their infrequency, it may
be stated that Louis found them in only four out
of forty-six cases; and Chomel relates, that out of
forty-two cases he did not meet with a single
example. Ulceration of the stomach may therefore be regarded as a comparatively rare lesion in
enteric fever.

Louis has drawn up a valuable résumé of these
lesions of the stomach, and by comparing the condition of the mucous membrane in persons who
have died from other acute diseases, he found the
same changes in this membrane as in enteric fever,
the proportions being very nearly equal in both.
For instance, softening with diminution in thickness was found in a fifth part of the cases of the

typhoid affection (enteric fever), and in a sixth
part of those of other equally acute diseases; ulcerations in the twelfth part of the former, and in
the twenty-fourth part of the latter; simple softening in a little less than a third of the cases of the
typhoid affection, and in a quarter of those of other
diseases. The mamellonated state was very nearly
in the same proportion. Finally, the mucous
membrane of the stomach was perfectly healthy
in two-sevenths of the patients who died of fever,
and in about a fifth of those who died of other
acute diseases. " Since the mucous membrane of
the stomach," he adds, " is not altered in all the
cases of this affection (enteric fever); since it is
found in a normal state in patients who die very
quickly, and in whom we cannot admit that the
disease, had it existed, could have disappeared
completely; since, also, in cases where one of
these lesions exists, it commences, as I have
shown, at a somewhat distant period from the
origin of the typhoid affection, it follows that the
typhus, ataxic, or putrid fever is no more a gastro-
enterite than a pneumonia is a gastro-pneumonia,
although we find the mucous membrane of the
stomach changed, in a more or less important
manner, in a great number of patients who die of
inflammation of the pulmonary tissue. So that
all we can deduce from the facts I have given—
and this conclusion is very important—is this, viz.,
that in every case in which an acute affection of

any kind gives rise to a febrile excitement lasting some time, the mucous membrane of the stomach becomes, at a period which varies according to the nature of the disease, the seat of a lesion, which lesion becomes more or less important according to the predisposition of the patient, while at times it accelerates the death, and is, in fact, in certain cases, the sole cause of it."

Louis has noticed another important point in connexion with this subject—viz., the relation which exists between the morbid states of the mucous membranes of the stomach and the symptoms observed during life. The conclusions he has deduced are—1. That in a considerable number of cases these various lesions, singly or combined, are found on examination after death, where there are no symptoms during life to indicate their existence. 2. That various gastric symptoms—epigastric uneasiness, either alone or accompanied with sickness—are not unfrequently observed in cases in which, on dissection, the mucous membrane of the stomach exhibits no appreciable deviation from its normal state. But, 3, in all cases in which there was epigastric disturbance, accompanied by bilious vomiting, more or less extensive change in the mucous membrane of the stomach is discovered after death. These conclusions have been confirmed by the observation of M. Chomel, in so far as relates to the indications afforded by symptoms, during life, of the lesions of the stomach; or, I should rather say, of the absence of characteristic symptoms.

The small intestines are generally more or less distended with air, and contain a moderate quantity of thin fluid tinged with bile, or, when there has been previous intestinal hæmorrhage, with blood.

The duodenum is generally unaltered, or, at all events, the changes are unimportant. Sometimes its mucous lining is of a deep-red colour, and more or less softened. In very rare cases the duodenal glands have been found swollen and superficially ulcerated.

It is almost unnecessary to repeat, that in certain lesions of the ileum—more especially of its lower third—we find the true anatomical characters of enteric fever. It is in this portion of the intestinal canal, where the agminated glands or Peyer's patches are chiefly localized, that the peculiar changes so characteristic of enteric fever take place. The diseased patches are readily recognised externally by the grey or brownish discoloration, or stains perceptible through the coats of the intestine, and by some degree of hardness when the diseased parts are pinched up with the finger and thumb.

With the view of making a minute inspection, the whole of the small intestine should be removed, and after carefully examining its outer or peritoneal surface, it should be slit open with the *gastro-*

tome—an instrument indispensably requisite for such investigations.

The first point to be noted is the colour of the mucous membrane. When the disease has been mild, or the patient has been destroyed by some accidental acute affection in its early stage, this membrane retains its natural pale pink-white colour. In cases in which the fever has been more severe, or further advanced, the membrane appears injected, or of a more or less deep-red tint, in patches of varying extent, but generally limited to the lower third of the ileum, and becoming more intense towards the cæcum. When the fever has been of still longer duration (twenty to thirty days), the colour is red-grey, or approaching to ash-grey, or sometimes it is yellow from bilious staining.

The membrane undergoes another change in a considerable proportion of cases — viz., in consistence. Louis found it softened in nine out of forty-three cases. The softening varies from a moderate to that extreme degree in which the membrane has little more than the consistence of mucus, so that it can be scraped off by the slightest application of the scalpel or nail. This may be partial or extensive, without any other alteration, or it may be combined with redness, or with thickening of the membrane, or with both. It is probable, however—and so thinks Louis—that this softened condition of the membrane may be connected with commencing decomposition.

In some cases, again, the membrane is found infiltrated with blood.

The changes which Peyer's patches undergo, and which constitute by far the most important point in the pathology of enteric fever, vary according to two circumstances—first, their proximity to the ileo-cæcal valve, and secondly, the duration of the fever. They are always most marked in the patches nearest this valve, and least so in those farthest from it; so that it appears that the altera· tions do not take place simultaneously but gradually or successively, first in those nearest to the cæcum, and then in those above, in the direction of the jejunum.

The earliest change consists in slight elevation or swelling (hypertrophy) of the mucous mem- brane covering the diseased patch, which appears also to affect, more or less, a large tract of the lower portion of the ileum. This is the probable condition of the intestine when the fever is mild and of short duration.

The next stage is that in which the deposit is effused into the tissue of Peyer's patches, the solitary glands, or into both, including the sub- mucous cellular tissue covering them. This typhous process may affect one or more of the follicles only, or an entire patch, and, according to the amount of the deposit, the mucous membrane becomes intimately blended with it, and more or less tense.

The swollen or infiltrated patches (which are placed opposite to the insertion of the mesentery) vary in size from an eighth or quarter of an inch to an inch and a half in diameter. Towards the lower end of the ileum, they occasionally occupy a space of several inches, and terminate on the border or margin of the cæval valve.

Louis has described under the term *plaques dures* a peculiar appearance in the patches which he found in about one-third of his cases (thirteen out of forty-six). He calls it hard in contradistinction to the soft kind of typhous infiltration, the difference arising from the transformation of the submucous cellular tissue into a firm homogeneous substance, having no organization, but friable when cut into. This appearance was observed in cases that had run a rapidly fatal course (from the eighth to the fifteenth day).

The patches undergo a still further change, that of ulceration—the pathological appearance most frequently observed in examination after death.*

The typhous deposit in this stage becomes transformed into a deep-yellow or brown slough, which in process of time is partially or completely detached from the patch, and ultimately thrown off, leaving a cavity or ulcer on the inner surface of the intestine, known by the name of the typhous ulcer. Sometimes the entire deposit is cast off at once;

---

* See Plate III.

in other cases only partially. In the former event, the shape and size of the ulcer correspond with those of the patch from which it has become detached. The ulcers may be either round or elliptical, their long diameter corresponding with the longitudinal axis of the intestine. When the slough has been only partially thrown off, the ulcer is smaller, and of an irregular shape. Several patches of ulceration may occasionally be seen grouped together.

In some cases the typhous product has apparently degenerated into a loose, vascular, fungous growth, traversed by streaks of extravasated blood, or entirely saturated with blood—a condition which is the frequent source of intestinal hæmorrhage.

A similar ulcerative process occurs in the solitary glands, but the changes take place somewhat later, and apparently advance less rapidly than in Peyer's patches. The form of the ulcer is round, not oval or elliptical.

Another circumstance connected with these typhous ulcers has been observed, viz., that when the typhous matter has been entirely removed from the diseased patches, there is no fresh deposit; so that if the excavation extend in depth, it is due to ulcerative absorption alone.

The following summary by Rokitansky indicates the progressive changes of this intestinal lesion :—

1. We find the lower third of the small intestine

to be the seat of the lesion, the number and size of the ulcers increasing as they advance towards the cæcal valve.

2. The form of the typhous ulcer, when it corresponds to the infiltration and detachment of a larger Peyerian patch, is elliptical; it is round when it corresponds to a solitary follicle, or a rounded patch, or to the partial detachment of a glandular plexus; it may also be irregular or sinuous, when corresponding to a partial detachment.

3. The size or circumference of the ulcer varies from that of a hemp-seed or pea to that of a half-crown.

4. The patches are placed opposite to the insertion of the mesentery, their long diameter being always parallel to the longitudinal axis of the intestine. The typhous ulcer never forms a zone; at least, Rokitansky has only seen this occur once in many hundred cases.

5. The base of the ulcer is formed by a delicate layer of submucous tissue, which covers the muscular coat, a well-defined fringe of mucous membrane forming the margin.

In regard to the progress of typhous ulcers, there seems no reason to doubt that the typhous deposit or infiltration affecting Peyer's patches may undergo the process of resolution. This happens in very mild cases in which the duration of the fever is short. There can be as little question that the ulcers cicatrize or heal. This reparative process may be seen in its different stages, when an opportunity is afforded of examining the intestine in cases in which death has unexpectedly occurred, either from some intercurrent local affection in the advanced stage of the fever, or during convalescence. When such an opportunity offers, some ulcers are observed to be superficial and not extensive, others deep and spreading, while in one or more patches (generally near the cæcum) the ulcers appear smooth and polished, and covered with a thin transparent pellicle, continuous with the submucous tissue around the ulceration—an indication that the process of healing, or cicatrization, is going on.

For this important result, however, two conditions are essential: the one, the termination of the local process—that is of the deposition of typhous matter, and the complete extinction of the typhous dyscrasia; the other, that the powers of the patient are adequate to withstand such a formidable local affection.

When such favourable circumstances concur, the following, according to Rokitansky, is the mode by which the process of cicatrization is effected. The fringe of mucous membrane which lies upon the base of the ulcer gradually connects itself, from without inwards, with the cellular tissue that invests the base, and, uniting with it, becomes paler and thinner. At the same time, the cellular layer becomes whiter and more dense, and finally

converted into a serous lamina, the circumference
of which is dovetailed between the muscular and
mucous coats. The margin of mucous membrane,
is beveled off in such a manner that the union is
imperceptible, while the line of union, as well as
the mucous membrane, is so thinned down, that,
at last, their villi appear to have been transferred
to the serous lamina. The edges finally unite at
one or more spots, and coalesce. When the healing
process is completed, instead of the typhous ulcer,
we find a slight depression on the inner surface of
the intestine, due to the thinning of the mucous
membrane, and the connexion with a thin cellular
layer of denser structure; or we observe a spot at
which the mucous membrane is more firmly attached and less movable, in the middle of which,
by oblique light, we may often discover a smooth
remainder of the serous lamina, of the size of a
millet-seed; or there may be a spot at which the
mucous membrane is more tense, void of plicæ or
folds, smooth, less vascular than the surrounding
portion, and particularly less villous. Such cica-
trices have occasionally been recognised thirty years
and even longer after the attack of fever.

Another singular and characteristic feature of
the typhous ulcer and its cicatrix is, that it never
induces diminution of the calibre of the intestine.

Again, in examining these ulcers from the
cæcum upwards, we find the healing process in
different stages. In some it has just begun, in

others it is going on partially, while in others it has been completed; so that the same course is apparently followed in the healing as in the development of this lesion, the cicatrization being either farthest advanced or completed in the patches nearest the cæcum, and in a proportionally less advanced state as we trace the patches upwards towards the jejunum.

In the destructive progress of the typhous ulcer, the mucous, areolar, and muscular coats may be successively laid bare, so that the floor of the ulcer is formed by the thin transparent peritoneum alone; and in some cases this membrane is perforated, so that the intestinal contents escape into the peritoneal cavity, giving rise to acute peritonitis.* The perforation is always found in the centre of an ulcerated patch, and most commonly in the lower third of the ileum; it is generally very small, not larger, perhaps, than a pin's head, unless there be a sloughing ulcer, in which case the aperture is proportionately large. There may be a single perforation only, or there may be two or more.

In some instances, the perforated portion forms an adhesion to a contiguous fold of intestine, or to some of the abdominal or pelvic viscera, and thus the escape of the contents of the bowels is pre-

---

* See Plate IV.—the delineation of a perforation in the centre of an ulcerated Peyer's patch, from a fatal case under my care in the Fever Hospital.

vented. I met with an instance of this kind which, however, ultimately proved fatal. The perforation was closed by a plug of omentum. I succeeded in obtaining a drawing of the parts by my late friend Sir Robert Carswell.

These perforations happen, as I have already remarked, not in the most severe cases of enteric fever, but more commonly when the disease is of moderate severity—when the ulceration is proceeding slowly but surely.

As to the period of the fever when this lesion takes place, it seldom happens before the middle period—towards the end of the second week; though it may occur at any time between the second and the eighth or ninth week. I have more than once known it to take place when the patient was supposed to be well advanced in convalescence, walking about and enjoying food.

The possibility of such an accident renders it imperative upon the medical attendant—fearless of the imputation of sordid motives—to watch with anxious care the period of convalescence in this dangerous disease. When, therefore, the convalescence is protracted, indicated by the general and popular phrase, that the patient does not get on— does not mend—does not gain strength—be assured that nature has not yet finished her work of cicatrization, and that an atonic ulcer has not healed, and is consequently pursuing a non-reparative course, which may end in erosion or perforation of the bowel, and the almost certain death of the patient. There may be even appearances of such improvement in the patient's general condition as to throw the most cautious and experienced practitioner off his guard. . The nutritive powers may be fairly active, the patient may be gaining flesh, the alvine evacuations natural, and even formed. But with all these advantageous circumstances, I have known intestinal perforation take place, and destroy life in a few hours. These, I admit, are exceptional cases, the perforation, in the great majority, happening in the earlier or more acute stage of the fever.

It is not easy to come to a certain conclusion as to the comparative frequency of this lesion. Louis met with it in eight out of fifty-five cases. In the last ten years there have been in the Fever Hospital, as far as I can trace, twenty examples of perforation out of the total admissions (1820). Of these, fifteen were males, and five females. The age of the youngest was eleven; of the oldest, forty-five. The preponderance in males is striking. In private practice, extending over more than thirty years, I have met with it eight times.

In tracing the large intestine, we find it generally distended with air. Its mucous surface is sometimes unaltered throughout; sometimes it is red, or of a grey colour, softened, and not unfrequently thickened in texture.

The solitary glands, which in the colon are small, are occasionally enlarged, and sometimes ulcerated, the ulcers, which are few in number, being small, with flattened edges, and the intervening mucous surface thickened. These ulcers are chiefly met with in the cæcum, to which they are often limited, though now and then they are scattered over the colon. They are rarely found in the rectum.

The ulcerative process in the colon is not, however, always confined to the follicles or crypts, but may extend to the intervening spaces where there are no follicles. It therefore appears that these ulcers in the colon are developed in two distinct tissues—in the solitary glands or crypts, and in the intervening mucous membrane; but whether they occur in the one or the other, they cicatrize or heal in the same manner as the ulcers in the small intestines, perhaps even more readily. I may observe, moreover, that these morbid appearances in the colon are more or less marked according to the length of time the fever has existed.

The mesenteric glands corresponding to the elliptical patches undergo changes in enteric fever as to size, consistence, and colour, in proportion to the extent of disease in the follicles, their proximity to the cæcal valve, and the duration of the fever. In patients who die between the eighth and fourteenth day they are simply increased in size, sometimes acquiring that of a bean or hazel-nut, and in those near the cæcum, they are even larger, the

tissue of the gland itself being elastic, soft, and of
a pale rose colour. The larger glands are sometimes so much softened as to be easily crushed by
squeezing. If death occur between the fifteenth
and twentieth days, they are swollen and soft, of a
red-brown colour, and on being cut into, the
typhous deposit may be seen in scattered points.
In patients dying between the twentieth and
thirtieth days, they are as large as pigeons' eggs,
and form a chain extending from the mesentery
attached to the ileum to the lumbar plexus.

The structure of the gland is converted into a
reddish-green medullary sort of matter, in which
there are occasionally deposits or extravasations of
blood; sometimes distinct fluctuation is perceptible, denoting collections of pus, either disseminated or in distinct cavities. In one case of abscess
of the mesenteric gland recorded by Louis, in which
death took place on the forty-ninth day, one of the
mesenteric glands was found converted almost entirely into pus, and the walls of the cavity in which
it was contained were so thin, that had the patient
lived a few days longer, it must inevitably have
opened into the abdominal cavity.

In regard to the progress of the mesenteric affection, Rokitansky states that as soon as the detachment of the typhous deposit in the intestine has
commenced, the swollen glands undergo a gradual
reduction in size, until they are restored to their
natural state; and Louis is inclined to think that

when they are found of a grey-blue or purple colour, not much enlarged or softened, a process of reparation had been progressing, analogous to what takes place when Peyer's patches are undergoing a healing or restorative action.

In some rare instances, enlargement and redness of the mesenteric glands have been found when Peyer's patches and the intervening mucous membrane were quite healthy, rendering it probable that the changes in these glands do not depend on antecedent intestinal lesions, but are due to the deposit or infiltration of typhous matter in their tissue. This, however, I have never met with.

The mesocolic glands undergo somewhat similar changes as to size, consistence, and colour, as the mesenteric, but to a much less extent, especially as to size. They are never found to contain pus.

I shall next allude to the changes found in other abdominal organs.

The spleen in its normal condition is of small size, so that percussion over the region it occupies fails to indicate its precise position, and it is only when it is considerably increased in volume, as in periodic fevers, and not unfrequently in both enteric and typhus fever, that its outline can be ascertained.

In enteric fever it undergoes, with few exceptions, marked changes; for in almost every case reported by Louis, it was more or less altered. He found it normal in four cases only out of the forty-six he examined. The examinations made at the Fever Hospital confirm this statement.

The changes observed in the spleen are—first, in volume; second, consistence; third, colour. As to volume or size, it may be only slightly increased, or two, three, or four times larger than natural. Its change as to consistence is less constant than its enlargement. In a certain proportion, a third perhaps, its consistence is unchanged; in about a sixth, it is moderately softened; and in about an equal number, the softening is considerable, so that it is easily torn. In some cases, it is little more than a pulpy or diffluent mass.

Louis found the size and consistence of the spleen much more frequently natural in patients who died after the thirtieth day of the fever than in those who died earlier. He also infers that, as he never met with a case in which death occurred between the eighth and twentieth days of the disease in which the spleen was not enlarged and softened, the changes commence probably at a very early period, and apparently exist in all patients; while in those cases in which no deviation from its natural condition is observed, it is possible that it had recovered its healthy state, in common with the other parts specially affected in the course of the fever.

It is less frequently altered in colour than in

volume and consistence; but it would appear that there is no constant relation between the alteration in colour and the size and consistence of the organ, as the same colour is found associated with every degree of both.

Among the organic lesions in enteric fever, the liver appears to escape almost entirely. It is seldom altered in size, the only change it has been observed occasionally to undergo being a degree of softening or friability, so that it is easily torn. This change, however, has been ascribed less to actual disease, than to the tendency to decomposition of the tissues after death; but though this condition is more generally met with in persons who die in the hot months, it is right to mention, that it has also been found in bodies examined in the cold season. It is also, in some instances, more or less congested—loaded with dark blood—but this may be regarded as accidental.

It does not appear that the kidneys undergo any changes that can be considered to be peculiar to enteric fever. In some bodies, they have been found slightly enlarged and softened, and somewhat darker in colour, than in the perfectly healthy state; but these conditions occur so frequently in persons who die of other acute diseases, that they cannot be considered to be peculiar to this fever. Of course, in examining the bodies of persons who have died of enteric fever, various lesions of the kidneys are occasionally found; but these have been of long standing, the fever having supervened on the renal malady, just as we often observe it supervening on other chronic diseases.

Of the salivary glands, the parotid is liable to inflammation and its consequences, the most common being the formation of matter in the cellular tissue surrounding the enlarged and indurated gland.

The submaxillary and sublingual glands become also occasionally enlarged from inflammation, either of their structure, or of the areolar tissue in which they are embedded.

With respect to the order of succession in which the several abdominal lesions commence and are developed, Dr. Bartlett states that is a matter not susceptible of very rigorous demonstration. Death almost never takes place in the disease before the termination of the first week, and not often so early as this. Still a careful study and comparison of the pathological appearances which are presented in cases of differing durations, will enable us to arrive at a reasonably certain approximation to the truth. There can be little doubt, he thinks, that one of the first—probably *the* first, pathological alteration which takes place in the solids, consists in the tumefaction of the elliptical plate or plates nearest to the ileo-cæcal valve. This tumefaction is accompanied or followed by other changes—an afflux of fluids, softening or the mucous coat, the hard yellow transformation of the submucous tissue, and finally, by ulceration; and these several lesions, taking place first in the plates nearest to the ileo-cæcal valve, gradually and successively extend to those which are farther removed from it. Contemporaneous, probably, or nearly so, with these alterations, are the reddening, enlargement, and softening of the mesenteric glands. The enlargement of the spleen and the diminution of its consistence occur also, there is good reason to think, in the early stages of the disease; and the same thing is also true, though less constantly, perhaps, of the softening of the other organs. The various pathological changes which are found in the gastro-intestinal mucous membrane begin, and are developed, it would seem, at uncertain and indefinite periods during the progress of the disease.

As to the changes in the brain and nervous system I may observe, that although the functions of this organ, as already pointed out, are, in the majority of cases, more or less seriously disturbed, neither its substance nor its investing membranes exhibit corresponding appreciable lesion; and even when morbid alterations are discovered, there does not appear, on close examination, that constant relation between the symptoms during life and the lesions found after death, that might, *a priori*, be expected.

But this does not apply to specific fevers only, but to cerebral affections in general, whether primary or secondary.

It will be convenient to consider the changes that take place within the cranium, first, in the membranes; and second, in the brain tissue.

The dura mater is almost invariably in its normal condition, idiopathic inflammation of this membrane, except after external injury, being, it is known, very rare.

The arachnoid membrane is, in some instances, opaque, sometimes slightly thickened; occasionally, a moderate amount of limpid or sometimes turbid serosity is effused into the sub-arachnoid tissue; or perhaps a small quantity of lymph, in the form of a delicate false membrane is found on the cerebral arachnoid. Effusion of blood into the arachnoid cavity—an occasional lesion in typhus—is rarely, if ever, observed in enteric fever.

The pia mater, in at least half the cases examined, is more or less injected, the injection being more apparent in those who die between the eighth and twentieth days of the disease.

Inflammation of the cerebral substance is never found after death in the most acute forms of fever. Indeed, well-marked inflammation of the entire brain tissue is a lesion so exceedingly rare, that its occurrence is doubted by many pathologists. Partial cerebritis is also to be regarded as an idiopathic disease only.

When the cerebral affection in enteric fever has been severe, the more common appearances are those of congestion—viz., injection of both the cortical and medullary matter—a decided rosy tint of the former and injection of the latter (shown by numerous red points), being apparent when a section of the brain is made. Louis found the cortical substance more or less of a rosy hue in seventeen out of the forty-six cases, and it was more frequent amongst those who died between the eighth and fifteenth days, than in those who died after this period. The medullary matter was injected more or less in all excepting seven, and this injection — generally in proportion to the colour of the cortical substance—was very marked in seven subjects, four of whom died before the fifteenth day; it was observed also in all who died between the eighth and fifteenth.

It thus appears, that injection of both portions of the brain (cortical and medullary) is almost invariably more pronounced, as well as constantly present, in patients who die in the early stage of unteric fever.

As to the consistence of the brain, it does not appear that there is any alteration in this respect from its normal condition. Softening is very rarely met with. Louis alludes to two cases only: in one, the softening was seated in the optic thalami, but there were no special symptoms which could be said to indicate the lesion; in the other, there was softening of the septum lucidum. In this case, however, the symptoms were well marked.

Serous effusion into the ventricles, more or less abundant (the average quantity being three drachms), is not uncommon, especially when the symptoms have been protracted.

The injection of the pia mater and of the medullary substance, the rose colour of the cortical substance, and the firmness of the entire mass, appear to be frequent and more marked, according to the more or less rapid manner in which the patient has died. On the contrary, when the fever has been of long duration, subarachnoid effusion with or without effusion of serous fluid into the ventricles, with diminished consistence of the brain substance, are the changes observed.*

The cerebellum in a large proportion of cases is in its normal state, even when there is preternatural vascularity of the brain. In the exceptional cases, and these are few, the changes are, moderate injection of the grey as well as of the white matter, with slight softening.

The pons Varolii is generally unaltered, the only deviation from its natural state (and this is rare) being slight injection of its structure, in cases in which there has been considerable vascularity of the brain substance and cerebellum. Louis observed it in one case only.

---

* Louis, op. cit.

In the medulla spinalis there are no lesions peculiar to enteric fever. Perhaps this portion of the nervous system has not as yet received the same rigid examination in fevers that has been bestowed on other organs. We have seen that other parts of the nervous system may be more or less changed from their normal condition, without giving indication by symptoms during life; and when we consider the delicately minute structure of the spinal cord, it is very probable that, if it become altered in the progress of fever, the morbid appearances may elude the search of the most expert anatomist.

In the respiratory organs, various pulmonary lesions are found after death.

Rokitansky ascribes the pulmonary changes to the peculiar typhous deposit (or, as he terms it, typhous process) which may be, he conceives, effused either on the epiglottis and larynx, the mucous membrane of the lungs and bronchial glands, or into their parenchyma, giving rise to laryngitis, bronchitis, or the so-called typho-pneumonia.

In general, the mucous membrane of the epiglottis and larynx is unaltered, but in some cases, it is red and somewhat swollen. In others, a thin exudation covers the epiglottis, and dips into the larynx (the laryngo-typhus of Rokitansky). It is frequently observed in the pharynx and larynx

simultaneously. Beneath this exudation, ulceration, superficial and limited, or deep-seated and of
some extent, may be found. We sometimes discover, too, œdematous or submucous infiltration of
the epiglottis and membrane covering the arytenoid
cartilages with or without muco-purulent fluid in
the laryngeal sacculi.

A dangerous, though less common, lesion of the
throat is the formation of an abscess in the pharynx, often in the immediate vicinity of the larynx,
and, though perhaps of no great extent, may be so
situated as to press on the epiglottis, and materially interfere with respiration. I have known,
in some instances, fatal results ensue from such
purulent formations. They often remain unde-
tected till revealed by examination after death.

The mucous membrane lining the bronchial
tubes, in a considerable proportion of cases, retains
its natural pale pink colour and delicate structure;
in others, there is diffused congestion, the membrane being dusky-red, swollen, and covered with
a blood-streaked, viscid secretion, especially ob-
servable in the lower lobes. Sometimes the secretion has a puriform appearance, indicating long
pre-existing bronchitis, on which the fever has probably supervened.

A singular opinion is entertained by Rokitansky,
that this morbid condition of the bronchial membrane may exist as a primary broncho-typhus, the
general disease—in fact, the whole phenomena of

fever—being originally localized, as it were, in this tissue alone, avoiding all other mucous surfaces, even that of the intestine, for which the typhous process, in general, shows the most decided preference. This speculative conjecture may or may not receive confirmation on further investigation, but it certainly is not at present considered even probable by British pathologists.

The bronchial glands occasionally undergo an alteration similar to that affecting the mesenteric: they are swollen, of a dark colour, and infiltrated with typhous matter. These glands (the bronchial), like the mesenteric, have been stated to undergo ulterior changes, that is, to become softened, and, with or without perforation of the adjacent mediastinum, to give rise to pleurisy. This I have never met with.

The pleural membrane of the lungs and thoracic parietes, as formerly noticed, not unfrequently becomes inflamed in enteric fever, so that after death the usual evidence of this secondary lesion is discovered. The pleurisy may have been either circumscribed or partial, with or without effusion, in moderate or considerable amount, and the exudation presenting the various characters it is known to assume in primary or idiopathic pleurisy. The effusion, however, generally shows that the pre-existing inflammation has been of a less sthenic character, the admixture of plastic exudation being in proportion to the more or less acute nature of the pleurisy. When of longer standing, the fluid, more or less abundant in quantity, assumes the puriform character—the condition also observed in those cases in which this lesion had existed in a latent form. It will be discovered, too, that the inflammatory process may not be limited to the pleural membrane, but, as in idiopathic pleurisy, to have spread to the subjacent pulmonary tissue.

When pulmonary inflammation has sprung up in the progress of enteric fever, the changes in the lung generally correspond with the nature, extent, and duration of the pre-existing affection.

The appearances in question are referable to one general condition—hyperæmia. There may be simply congestion or engorgement, chiefly of the posterior and inferior portions of the lung, in consequence of the gravitation of the blood to the more dependent parts. When a piece of such congested lung is cut into, it exhibits a dark-red colour; the lung tissue is swollen and somewhat dense, and filled with a frothy, sanguineous, or sero-sanguineous secretion, and according to the degree in which these changes have taken place, the portions, when plunged into water, may or may not sink.

When the pulmonary hyperæmia has been of a more acute character, we find other changes in the lung. They, too, are usually observed in the inferior portions, the upper and central being less frequently the seat of pneumonic inflammation.

On examining the inflamed portion, it exhibits a violet-red colour, is more or less solid, non-crepitant, and sinks in water. Or perhaps, from the cellular or areolar tissue of the lung having been blocked up by the exuded products, the diseased portions have acquired a granular appearance, resembling the structure of the liver—in other words, have become hepatized. But the more advanced stage of pulmonary hyperæmia (grey hepatization) or the ulterior change (purulent infiltration) is very rarely observed in the secondary pneumonia of enteric fever.

These lesions may involve the whole or a greater portion of one lung (lobar pneumonia), or they may be limited to individual lobules, the intervening lung structure being normal (lobular pneumonia).

Again, among the pulmonary lesions or secondary affections of this form of fever (and peculiar to it alone), but differing in some respects from the changes induced in the lung tissue by common non-specific pulmonary hyperæmia, is that distinguished by the term splenization, from its resemblance, in some respects, to the structure of the spleen. It has also been called carnification. It was first, I believe, pointed out by Louis, and subsequently recognised by Dr. Bartlett as a lesion frequently observed in the enteric fever of the United States, and has not escaped the notice of British physicians. It has been well described by

Rokitansky as a higher degree of hyperæmia, in which the parenchyma of the lung becomes saturated, as it were, with blood, so that it assumes a dark-red or slate-purple colour.

When the affected lung is minutely examined, it has a mottled appearance externally, in which may be seen dark-coloured patches or portions, of varying size, of a somewhat dense or solid structure, easily torn, but still crepitous. If the affected portions are cut into, dark fluid blood escapes, but when plunged into water, they float. In a more advanced stage of this lesion, the diseased patches increase in density, and cease to crepitate or to float in water, and when deep incisions are made, very little blood can be squeezed from the cut surfaces. When the air is entirely removed, with the exception of the gaping mouths of blood-vessels, no trace of pulmonary structure is discernible, but only a tough, disorganized, friable mass, very unlike the granular appearance of consolidation in ordinary pneumonia.

The term pulmonary hypostasis has been applied to a particular condition of the lung in which a passive congestion—a stasis—takes place in the posterior and inferior portions, in persons who have been long confined to bed by protracted diseases. It has been occasionally detected when fevers and other acute maladies have been of unusually long duration. The course of this hypostatic congestion is slow and inactive, and it has

K

been supposed, in exhausted subjects, to have sometimes laid the foundation of latent inflammation of
the lungs.

Cadaveric pulmonary hyperæmia—a post-mortem hyperæmia of the lungs—should be distinguished from the hypostatic congestion just noticed,
which always takes place during life. In consequence of prolonged imbibition in these cases, the
pleural membrane not only becomes discoloured,
but a certain quantity of sanguineous serosity occasionally makes its way into the pleural sac.

I may here allude to the notion sometimes entertained, that pulmonary tuberculosis is an occasional sequence of enteric fever. This idea by no
means accords either with my own experience, or
with that of physicians who have had large opportunities of observation. It has originated in tracing
the changes that take place in the air-cells in pneumonia, which, as is well known, consist in inflammation of their structure, and the gradual oblite-
ration they undergo by inflammatory products.
When the cells are filled with such plastic material, the diseased mass becomes solidified, and the
cut surface has a granular appearance. These
changes, which are only the advanced stage of
vesicular pneumonia, constitute the lesion described
as Bayle's pulmonary granulations, which from its
supposed resemblance, in some external characters,
to the pulmonary parenchyma when infiltrated with

tubercular deposit, has been mistaken for it, and in this way may be explained those cases of pulmonary tuberculosis which have been said to follow fevers.

The organs of circulation seldom undergo much change in enteric fever.

The pericardium is generally, almost invariably, in its natural state. It occasionally contains an inconsiderable amount of serous fluid, but without trace of pre-existing inflammation.

In about one half, or perhaps in a still larger proportion of the cases examined, the muscular structure of the heart is normal; and when there is deviation from the natural state, the chief alteration is softening, in which the walls lose their natural cohesion or strength, are diminished in thickness, and easily lacerated. When removed from the body, the organ is somewhat flaccid, and when cut into, its tissue is pale and soft. Sometimes the internal lining is stained dark-red, an appearance which should not be mistaken for the effect of inflammation.

In a memoir read before the Academy of Medicine in 1830, Andral threw out the suggestion that this cardiac softening might be regarded as the result of commencing cadaveric decomposition, favoured by circumstances which it is difficult to appreciate. This view, however, was combated by Louis, who maintained that the softening probably

K 2

commences before death, and is therefore a lesion, not a post-mortem phenomenon: and more especially so, when it is remembered that the frequency and severity of the softening are more marked according as the fever is more early fatal. Thus he found the heart softened in nearly half of those patients who died between the eighth and the twentieth days of the fever; in a third of those who died from the twentieth to the thirtieth; and in a somewhat smaller proportion amongst those who died later: and he adds, that had he included cases omitted, in which the softening was less marked, the proportion would have been much larger. He brings forward another cogent argument—that no similar lesion (softening) was found in the other muscles, which retained their natural colour and consistence.

In the aorta, the only deviation from the normal state occasionally met with, is more or less redness of the inner membrane, (in some cases diffused generally, in others in patches,) spreading some way into the descending aorta and the large arterial branches. It frequently extends to the middle coat, and is generally accompanied with cardiac softening; and it has been both more frequently observed, and more marked, in patients who die in the early than in the advanced periods of the fever. The notion, however, that this redness (of the internal lining of the aorta) is to be ascribed to hyperæmic changes, has been exploded as unten-

able, the more satisfactory explanation being, that
it is merely staining from imbibition.

The blood contained in the chambers of the
heart varies according to the condition of this
organ. When its consistence is normal, fibrinous
clots, more or less firm, are found. When it is
softened, the blood is defibrinated, and in soft semi-
fluid masses in the auricles and ventricles; when
the softening is extreme, the blood is fluid, in small
quantity, and mixed with air-bubbles.

It is almost unnecessary to observe, that of the
various lesions described some only are to be considered as the essential elements of enteric fever;
indeed, but a small portion can be viewed in this
light. It is, however, important to determine, as
far as possible, the bearing which, separately or combined, they have on the pathology of this disease;
and although much has been lately accomplished in
elucidating the changes that take place in the several
organs in its progress, this question is one of great
difficulty, so complex, indeed, that it cannot be
said to be susceptible of precise adjustment in the
present state of our knowledge. Still a great step
has been gained in determining the true or essential
pathological element of this fever; and it may
be reasonably assumed, from the fact that two important lesions are invariably present—viz., the
alterations in Peyer's patches, and in the corresponding glands of the mesentery—that these two

lesions constitute its true anatomical character. In this, almost everyone who has investigated the subject since it was first pointed out, and I may say settled, by the investigations of Louis, unhesitatingly coincides.

The other lesions, when present, are to be regarded as secondary or accidental only. Some of them, however, which are of very frequent occurrence—the alterations in the mucous membrane of the stomach, in the spleen, and in the lung for example—have certainly great influence over, and often determine the issue of the disease. And it would appear that a favourable or unfavourable result is in many cases to be ascribed as much to their coexistence, as to the extent or severity of the primary or essential intestinal and mesenteric lesions.

Another point of unquestionable importance to a right understanding of the pathology of enteric fever—if it could be settled—is the relation which the abdominal lesions bear to the phenomena of the disease. What inferences, for example, are to be deduced from the alterations in Peyer's patches? Are we justified in considering this lesion as the cause of the whole febrile phenomena? In other words, is enteric fever merely a gastro-enteritis, or a dothin-enteritis, just as inflammation of the mucous membrane of the colon is regarded as the cause of the morbid condition included in the term dysentery? Modern pathologists have come to the conclusion that this is not a correct mode.

of viewing the question, and, as I have repeatedly
stated, the intestinal lesion and changes in the
mesenteric glands are not to be viewed as the
cause of the febrile phenomena, but as only a
part of the pathological elements of the fever,
the whole being the result of some morbific
agent, of the nature, sources, and modus operandi
of which, we are as yet entirely ignorant. The
lesion of the patches certainly bears to enteric
fever the same relation as does the eruption of
small-pox, of measles, or of scarlet fever to the
co-existing fever, in which no one regards the
particular cutaneous rash to be the cause or source
of the febrile affection.

TYPHUS FEVER.

Nature of typhus—invasion and progress of the symptoms—typhus eruption—cerebral symptoms—muscular system—organs of the senses—the pulse—temperature—digestive and excretory system —secondary affections—of the brain—of the pulmonary organs— bronchitis — pneumonia — pulmonary gangrene — pleurisy — the heart—duration—crisis and critical days—pathological anatomy—absence of lesions in the alimentary canal—changes in the blood—diagnosis.

I HAVE dwelt at some length on the diagnostic characters of enteric fever, showing that they consist, first, in certain alterations in the agminated glands (Peyer's patches) and corresponding glands of the mesentery, by which lesions this form of fever is distinguished, not only from typhus, but from all acute diseases; and second, in a peculiar rose-coloured, spotted rash, the distinctive features of which have been pointed out.

Typhus fever, on the other hand, is characterized by the absence of specific local affection, if we except the mulberry eruption, by which it is, if not invariably, at all events in a large majority of cases, accompanied. When lesions are found after death, they are due to secondary, or intercurrent diseases (complications) which have arisen in the progress of the disease.

The importance of being enabled to distinguish these two forms of fever has been repeatedly insisted on—not as a mere pathological nicety, but as all-important in a therapeutical point of view, since they differ, not only in symptoms, duration, and fatality, but in many respects, in treatment. In short, typhus fever is pre-eminently the type of a blood disease, the fever poison acting primarily on the blood, and secondarily on individual organs or parts, though after death, there may be no trace of structural changes, except such as have resulted from intercurrent affections; differing in this respect from uncomplicated enteric fever, in which, as we have seen, there are invariably well-marked special lesions.

It appears very probable, too, that the two fevers differ in their etiology. Typhus often prevails epidemically, from unknown causes, besides being the almost invariable accompaniment of famine and destitution, and spreading rapidly in over-crowded and ill-ventilated dwellings; and when once engendered, from whatever cause, is propagated by contagion, regardless of age, sex, or individual circumstances.

Enteric fever, on the other hand, appears to have less of an epidemic, more of a local origin; and though its causes are very obscure, it is supposed to be connected with ill-constructed dwellings and malarious exhalations.

Typhus therefore differs from enteric fever in the absence of primary local disease, in etiology, and, as we shall presently see, in special symptoms.

But before detailing the symptoms by which typhus fever is characterized, I may remark, that the susceptibility to the influence of the poison, however generated, varies, and hence the difference observed in the accession or mode of invasion of the disease. Some experience sudden indisposition after exposure to an infected atmosphere, or contact with a person affected with typhus. For example, when nurses are attacked after their ministrations to fever patients, they are often able to connect their illness with services rendered to an individual patient; and in my own case, I remember, to this day, the feeling of being struck with fever, after leaving a particular house, during the epidemic which prevailed at Edinburgh in 1817-18.

But more commonly, there is a period of incubation, of longer or shorter duration, in which various undefined sensations—irregular chills or shivering, nausea, disrelish of food, thirst, giddiness, perhaps slight pain in the head, somewhat accelerated pulse, with languor and muscular aching—are felt, for days perhaps, before the fever can be said, in popular language, to be "formed." As the disease progresses, these sensations become intensified; there is visible change in the countenance, the patient is agitated and restless, while the symptoms of general disturbance alluded to are more pronounced, and recur in paroxysms of uncertain duration, but chiefly towards evening, or in the night, which is spent in wakefulness, or disturbed sleep.

These symptoms, however, are common to other forms of fever, and even to acute diseases in general, so that it is by their progress and further development, that we are enabled to form our judgment as to the precise nature of the disease they indicate.

A prominent feature is the peculiar distinctive eruption (to which I have already cursorily drawn your attention) known as the typhus rash, or, from its resemblance to the efflorescence of measles, it has been called mulberry or morbillous. When this eruption is present, it indicates at once to the experienced eye, this form of fever.

A comparison of this rash (a representation of which is given in Plate II.), with the rose-spotted eruption of enteric fever, will enable you at once to recognise the one from the other.

Let me draw your attention more particularly to this characteristic eruption. It is preceded and accompanied by a fainter efflorescence, on which darker red spots are scattered, giving the portions of skin a mottled appearance. It is visible from the fifth to the seventh day of the fever, the spots of which it is composed varying in size from a point to three or four lines in diameter, and having an irregular outline. Sometimes they are few in number, but more commonly they are numerous, the larger spots being formed by the coalescence of smaller ones. At first they have a dusky pinkish hue, fade or partially disappear on pressure, and when the finger is passed over, they give the sensation of being slightly elevated. After a day or two, they assume somewhat of a brick-dust colour, scarcely fade on pressure, the tint gradually becoming more and more dark, and not disappearing when even firmly pressed. Another distinctive peculiarity is, that each patch, or cluster of spots, remains from its first appearance till the termination of the disease, thus differing from the rash of enteric fever, in which, as has been pointed out, the spots not only disappear on pressure, but last for a few days only, and are succeeded by a fresh eruption, so long as the fever continues.

Dr. Stewart noted the period of the fever when the typhus eruption became visible in fifty-two cases. He observed that in twenty-nine—more than one-half—it appeared on the fifth or sixth day, and in three-fourths, from the fourth to the seventh. In forty-eight (four having died before the eruption began to recede), the eruption began to decline at different periods from the eighth to the nineteenth day, its average duration being eleven days and a half.

In some patients, this rash, though well developed, is less prominently marked. Many of the individual spots are pale and undefined, and though grouped in patches, they are irregularly scattered, so that the portions of the skin have a faint dingy appearance. Again, there is often considerable modification in the colour, as well as in the copiousness of the efflorescence, the abun-

dance, and especially the deep colour of the eruption, being as a general rule, in proportion to the
severity of the symptoms.

This has been repeatedly noted in the records of
epidemic fever. For example, in the Glasgow
epidemic, so well and faithfully recorded by Dr.
Stewart, it was observed, that when the cases
were few, and the fever mild (the death-rate being
one in ten or twelve only), the eruption, if present at all, was scarcely visible; but when the
cases became more numerous, as well as more
severe (the per-centage mortality being higher—
from one in eight and a-half to one in six and
a-half), the eruption was proportionally developed.
In the Edinburgh epidemic (1838-9) the mortality
of cases with scanty eruption was one in eight;
when the rash was abundant it was one in four; the
duration of the fever being also between one and
three days longer in the one than in the other.

In regard to the regions of the body on which
it appears, when abundant, it covers both the
anterior and posterior aspect of the trunk, as well
as the limbs, coming out first on the trunk, and
spreading to the extremities, being rarely, however,
seen on the face. When less abundant, it is limited
to the chest and abdomen, and, as a general rule, it
seems to reach its maximum within the first three
or four days after it becomes visible.

Is this distinctive rash, it may be inquired,
always present in typhus? It is occasionally

absent. And here statistical facts are valuable in determining the relative proportion of cases in which it was not found. Of one hundred and fifty-two cases, of which accurate notes were made in the journals of the Fever Hospital by Dr. Jenner, it was present in one hundred and thirty-six, and absent in sixteen. It was generally observed, too, that when the rash was absent, the fever was mild and of short duration, unless some intercurrent or secondary affection occurred, and altered the character of the disease. Of the sixteen cases in which it was absent, thirteen were between the ages of four and fifteen, and the remaining three, between sixteen and twenty-two. It also appeared, that of the whole number (one hundred and fifty-two), seventy-six (exactly a half) were more than twenty-two years old, and twenty-one between the ages of sixteen and twenty-two inclusive. It was therefore present in every individual more than twenty-two years of age; it was absent in three whose ages varied between sixteen and twenty-two; and absent also in thirteen (or about a fourth) of those fifteen years of age and under, and of these thirteen, there was only one whose age exceeded thirteen.

The inference to be drawn from the facts stated is, that of one hundred individuals of all ages with typhus fever, we may expect, as a general rule, to find the rash absent in twenty-five (or one-fourth) of those under puberty; from fourteen (or about one-seventh) of those under manhood; but from none above twenty-two years of age. So much for the statistics of this rash.

As it is important to keep in view the distinguishing characteristics of the typhus eruption, let me restate them.

1. It appears from the fifth to the seventh day of the fever, and consists of a congeries of spots, which cohere in the form of patches on the trunk and limbs (seldom on the face), of a brownish or mulberry hue, giving, when well developed, a diffused or mottled appearance to those parts of the skin on which it comes out ; fading, but not entirely disappearing, on pressure; differing in those respects from the bright rose spots of enteric fever, which are isolated or distinct, and entirely vanish on firm pressure. 2. The typhus eruption is permanent or persistent—that is, it continues from its first appearance until the fever terminates; unlike the rose-spotted rash peculiar to enteric fever, which, after running through successive changes, disappears in a few days, and is succeeded by a fresh eruption of spots, which go through the same changes, crop after crop, until the cessation of the fever. 3. Though this eruption may be discovered by careful examination in a large proportion of cases, it is occasionally absent, its absence, however, being more commonly observed in children than in adults. Its presence or absence, moreover, has an

important bearing on the character of the fever—its absence generally indicating a mild disease.

Of the general symptoms of typhus, those referable to the nervous system are the more prominent. Headache is one of the most constant in the early stage, indeed it is rarely absent. It may be so slight as not to attract the patient's notice; but more commonly, for the first week or ten days it is severe and persistent, after which it abates gradually, and finally disappears towards the end of the second week; or, should it continue longer, it comes on at intervals only—in the evening perhaps, or during an exacerbation of the fever.

The intelligence is seldom disturbed in the milder cases. In the more severe, there is transient confusion, soon lapsing into delirium, perceptible at first only on awaking, or in the night, but becoming more constant as the fever advances. As to the character of the delirium, it is sometimes noisy and violent (typhomania); more often, low and muttering, with tendency to somnolence; in other cases, it partakes of the character of delirium tremens—the form usually observed in spirit-drinkers and careworn individuals.

This symptom (delirium) is rarely noticed before the seventh or eighth day. It may appear, however, earlier or later; but at whatever period it comes on, it continues till the termination of the disease.

Stupor or somnolence, in a greater or less degree, is seldom absent. It may occur with or without previous delirium, supervene in the early or not until an advanced period of the fever, and gradually disappear, or slowly increase until death ensue. As to degree, there may be every modification from unnatural tranquillity, transient somnolence, to insensibility or deep coma. In the worst cases, generally fatal, the patient lies perfectly unconscious, with the eyelids open as if awake—a condition to which the term *coma vigil* has been applied.

It may be observed too, that in typhus, the brain symptoms are not only more prominently marked, but appear earlier than in enteric fever.

The whole muscular system, evidently influenced by the condition of the brain, undergoes early and marked changes in typhus—much more marked than in enteric fever, the general prostration being greater, while the peculiar alteration in the physiognomy is often so striking as to be in some measure a guide in our diagnosis. Many persons previously robust, become feeble and powerless to a degree that would hardly be expected, so as to necessitate confinement at once to bed. This muscular weakness increases with the progress of the fever, and in severe cases the patient becomes utterly helpless, and often scarcely able to turn in bed without aid. In the still more grave, other symptoms referable to the muscular system appear — sometimes early, but more generally in the advanced periods of the fever—in the form of muscular tremor, spasmodic twitchings of the tendons of the wrists (carphology), or the muscles of the face, or of the diaphragm, giving rise to hiccup. In some cases there is general muscular tremor, resembling a severe chill or rigor.

In typhus, the organs of sense become more or less affected. The sense of hearing is often impaired, deafness in various degrees coming on and lasting so long as the fever continues. Permanent imperfection, however, is very rare.

In cases of moderate severity, the organ of vision is little affected: in the more grave, with severe cerebral affection, the eyes appear dull, with more or less dark or dusky suffusion of the conjunctiva, and pain on exposure to brilliant light. When there is coma, they are still more heavy, and the suffusion deeper, there being often also inability to close the eyelids. Towards the termination, patients attempt to catch imaginary objects, or to pick the bedclothes, under the false notion that they are covered with spots. Occasionally they complain of seeing double, or through a mist.

The sense of taste, too, may be either impaired or perverted, or even entirely lost; hence articles of food are refused, under the idea that they are ill prepared, or have lost their usual flavour.

The sense of smell may be diminished or, like that of taste, entirely lost.

In mild typhus, there is little heat of skin at first, but as the fever proceeds, it becomes hot, often pungent, and accompanied with dryness, the temperature, as ascertained by the thermometer, ranging from 98° to 108° (Fahrenheit). In some patients the skin emits a disagreeable odour, so peculiar as to be readily recognised, though difficult if not impossible to describe.

The pulse is generally frequent, soft, easily compressed, and often irregular. The heart may partake of the general muscular weakness, so that the sounds are scarcely audible. To this cardiac weakness is to be ascribed the tendency to syncope when patients attempt to assume the erect posture too early, as well as the occasional fatal accidents when, feeble and prostrate, they are permitted to pass their evacuations in the night-chair.

The digestive system, which is so prominently affected in enteric fever, is little if at all disturbed in typhus; so that the absence of gastric or intestinal irritation constitutes an important point in diagnosis. Sickness or vomiting may be said to be rare, nor does the abdomen exhibit externally anything abnormal, there being neither unnatural prominence nor tympanitic distension. Nor in regard to the condition of the bowels is there, in general, any deviation from the ordinary habit of the individual: at all events, if there be a tendency to relaxation, it is accidental and temporary, while the evacuations are feculent, not watery, as in enteric fever. It may therefore be considered an axiom, that spontaneous diarrhœa, which is a very prominent symptom of enteric fever, is very rarely observed in ordinary typhus.

Again, intestinal hæmorrhage, so frequent in the former, rarely occurs in the latter; and if it do occur, depends either on congestion of the mucous membrane of the colon, or on hæmorrhoids connected with a gorged condition of the portal system.

The tongue in the beginning of typhus is covered with thin white mucus, which in its progress becomes more or less brown, thick and tenacious. In the more grave cases, this incrustation is still darker, becoming at length almost black or fuliginous, while the tongue itself is shrivelled and fissured, and sometimes coated with blood.

The teeth and lips too are incrusted with a similar exudation of dry, offensive mucus, or with bloody sordes, which evidently issues from the mucous surfaces within the mouth—a local hæmorrhage, in fact,—as blood-discs may be discovered on placing a portion of this sordes under the microscope.

With this dry parched state of the tongue and fauces, there is often some difficulty in thrusting it forward, or in swallowing, especially when the first efforts are made.

The appetite or desire for food, especially solids, is considerably impaired, and the thirst is generally in proportion to the degree or intensity of the

general symptoms. Some patients exhibit little desire for fluids throughout the progress of the fever,
but these are exceptional cases, and it has always
been regarded as an unfavourable symptom if the
patient do not complain of thirst when the mouth
and fauces are very dry, denoting great insensibility.

I may just advert to the bladder, not that it exhibits any deviation from the natural state in typhus, but to draw your attention to the importance
of daily careful inquiry as to its expulsive powers.
In most cases, especially where there is little brain
disturbance, the urine is passed as in ordinary
health; but sometimes there is insensible or in-
voluntary dribbling, so as to lead to the impression
that there is no accumulation in the bladder. If
more minute inquiry be instituted, especially by
percussion over the space occupied by this organ,
the distended bladder is often detected, requiring
the use of the catheter. It is a safe rule, therefore, to inquire, at least once a day, as to the state
of this organ, in order to be satisfied that there
is no inconvenient distension.

But bear in mind that the secreting powers of
the kidney may become suspended—suppression of
urine,—which is a serious matter, from the possibility of urea becoming mixed with the blood, and
giving rise to the phenomena of uræmic poisoning.

That changes, too, in the physical and chemical
characters of the urine do take place in fevers, is

evident from the various alterations it exhibits in their progress. On these, however, I do not intend to enter, nor am I aware that, even in the present advanced state of chemical science, any pathological or practical conclusions have been accepted, in explanation of the causes of the various changes which the urine undergoes in febrile diseases.

In typhus, as well as in enteric fever, we must be prepared for the supervention of secondary affections, and certainly those of the brain are the more common, and often give the practitioner great anxiety. We have seen that from the beginning, there is almost invariably pain in the head, more or less constant, which however passes off seldom persisting after the second week. But in other cases, as the fever proceeds, the cerebral symptoms predominate; there is more deep-seated headache, accompanied by flushing and heat of scalp, heaviness of the eyes, with conjunctival sufffusion, and aversion to light, persistent delirium, and in still graver cases, somnolence, gradually lapsing into coma, more or less deep.

Such symptoms show that the nervous system is severely implicated in the general commotion, and may be looked for in persons who have previously sustained mental shocks, as well as in the intemperate, and in those who have indulged freely in tobacco or opium. I remember being much perplexed with the strange anomalous nervous symptoms exhibited by a man under my care in the hospital many years ago. The patient at length, in a state of half-drunken delirium, implored me to give him opium, and acting on this hint, I prescribed two grains of solid opium at short intervals, which had the effect of calming the nervous symptoms, and ultimately inducing sleep. This man admitted that he had for years been in the habit of taking from twenty to thirty grains of opium daily, and I thought it the best practice to give him this remedy, but in greatly diminished quantities, until convalescence was established. He appeared to get rapidly better after the first day the opium was prescribed.

In children, brain affections are by far the most frequent of the secondary lesions, in fever as well as in all acute diseases. It is therefore especially necessary to watch the first indications of the brain becoming prominently affected. Hence increased restlessness or impatience, flushing, more constant complaints of pain in the head, intolerance of light and sounds, wakefulness, or delirium, should at once lead the practitioner to watch closely the condition of the brain. When there is drowsiness, with or without intervals of sleep, lapsing gradually into more or less deep coma, it is very difficult if not impossible, at this stage of the disease, to decide on its precise nature, whether it be the advanced stage of tubercular meningitis, or typhus with severe cerebral affection. The early history, if it can be ascertained from a reliable source, and the presence or absence of the characteristic mottled or mulberry rash, tend to clear up the doubt. Nor is the diagnosis unimportant, either in reference to the prognosis or treatment: the one, tubercular meningitis, being surely fatal ; the other, typhus with acute brain affection, although hazardous, not entirely hopeless.

There is another point that requires notice. This secondary brain affection is liable to obscure, or render latent, disease in another part—the chest, for example; and hence the necessity for frequent auscultation of the pulmonary organs when there is more than usual cerebral disturbance, in this as in other forms of fever.

As to secondary pulmonary affections, I may observe that changes in the lungs are more commonly found, after death, in typhus than in enteric fever; indeed, pulmonary congestion is almost invariably present in typhus.

The most common secondary affection of the lung is bronchitis, which in the slighter cases is limited to the larger tubes, and may be at once recognised by irritative cough, and loud coarse rhonchus under the clavicles. The bronchial affection may continue thus limited, but more commonly it spreads to the smaller ramifications, ultimately inducing secondary capillary bronchitis, which may be recognised by increase of the general fever, frequent paroxysms of cough, dyspnœa, and præcordial constriction. When it assumes a latent form, the general symptoms are absent, and it can only be detected by auscultation, which reveals its possible extent and intensity. The percussion note over a considerable portion of one lung is less clear than natural, but the resonance, though diminished, never amounts to more than slight dulness; and as to the stethoscopic signs, there may be every variety of sonorous, sibilant, and mucous râles, which sometimes are scarcely distinguishable from the crepitation of pneumonia. Observe, too, posture has often a remarkable influence on the physical signs—a circumstance rarely observed in primary bronchitis. For example, if the patient has continued for several hours on either side or on the back without change, the lung will be not only more dull over the side on which he has been placed, but the respiratory murmurs are more loud and less vesicular. In some cases, again, even when this secondary bronchitis exists to a considerable degree, the morbid sounds are very indistinct—perhaps entirely absent. The reason (according to Dr. Stokes) appears to be, that the finer ramifications of the bronchial tubes are so turgid, that during ordinary breathing the air does not enter them with sufficient force to produce a sound, until the patient makes—which he is not always able to do—a forced inspiration, when the sounds become audible. If the patient recover, as the lung improves, the râle during ordinary breathing becomes more distinct and constant, the increase of râle during ordinary breathing indicating a decrease of this bronchitic affection.

It is always difficult, as I have said, to distinguish secondary capillary bronchitis from pneumonia. Indeed it often runs into a low form of inflammation of the pulmonary parenchyma, the whole or greater portion of one lung exhibiting dulness on percussion, with the characteristic fine crepitation of pneumonia. Still the dulness is less complete than in ordinary pneumonia, the pulmonary tissue being in a state of congestion rather than of consolidation.

Intercurrent pneumonia may arise either in this way, or it may become openly developed and recognised by its physical signs. It may be limited to a portion, or involve the whole of one lung; or it may involve both lungs—this secondary double pneumonia proving, with rare exceptions, rapidly fatal. Sometimes it creeps on slowly and insidiously, assuming, so far as its general signs are concerned, a latent character, there being neither cough, accelerated breathing, nor local uneasiness to indicate its existence, which is only discovered by the aid of physical diagnosis, when its peculiar crackling sound is revealed.

The so-called hypostatic bronchitis, occasionally running into pneumonia, is a peculiar condition of the lung, supposed to be produced by the gra-

vitation of the blood in the posterior and depending parts of the pulmonary structure, from
long confinement on the back, in consequence of
the patient being unable to turn in bed. Some
pathologists, on the other hand, are inclined to think
that it takes place immediately before death, when
the act of dying has been protracted. But whether
it takes place immediately, or some time previous
to dissolution, it arises in the mechanical way I
have explained.

Pulmonary gangrene occasionally occurs as the
sequence of secondary pneumonia. It is recognised
by the signs indicative of pulmonary excavation—
bubbling sound of the breathing at the apex of the
lung, with peculiar fœtid gangrenous odour of the
expectoration and breath. Sometimes the ex-
pectorated secretion only is fetid, the breath being
inodorous, except when the patient coughs. It is
always fatal, according to my experience.

Pleurisy is much less common in typhus than in
enteric fever, in which latter form there appears to
be a proclivity to serious inflammations, while in
typhus local congestions predominate. Still, now
and then we meet with a low form of pleurisy
springing up in the progress of typhus, with or
without pneumonia, and leaving after death traces
of its existence in the effusion of lymph of limited
extent and extreme tenuity, from the diminished
amount of fibrin in the blood.

The heart does not appear to undergo any special

lesion in typhus. It partakes, however, of the general affection of the solids, so that its contractions, and consequently its sounds, become more or less enfeebled in proportion to the general depression. This alteration is not to be detected in the first stage of the fever, generally not until the end of the second week, when the first or systolic sound becomes more feeble, less prolonged, and more abrupt, so as to be scarcely distinguished from the second or diastolic. In extreme cases, indeed, this sound is nearly extinguished. Hence the importance of ascertaining the power or quality of the heart's contractions, rather than its frequency or number of pulsations; for it has been observed, that when the first sound is so feeble as to be nearly inaudible, the patient seldom recovers. In this feeble condition of the heart the pulse at the wrist participates, being soft, easily compressed, generally rapid, and sometimes irregular; but let me repeat, it is always better to ascertain the force of the circulation by examining the condition of the heart itself, than the pulse in any of the arteries.

The duration of typhus is considerably shorter than that of enteric fever, and when it proves fatal death takes place, in general, at an earlier period. If we trace the amendment, or turn, as it is called in popular language, it will be observed that this happens generally between the tenth and sixteenth day, but the average duration may be assumed to be fourteen or fifteen days. In mild cases, it is short; in the more severe, it is protracted, seldom, however, exceeding twenty-one days. It will be found, too, by comparing its duration at different ages, that it is shorter in young persons—that is, under the age of puberty —than in adults.

These remarks apply, of course, to simple or uncomplicated typhus, for when intercurrent affections spring up, it must be apparent that its duration cannot be defined.

The approach of convalescence is indicated by a gradual decline in the more characteristic symptoms. The delirium, often one of the most prominent, first disappears, and the patient recovers, in some degree, the knowledge which, for a time, had been lost. The memory, however, though much improved, is evidently defective; hence the difficulty in recalling to mind circumstances that had occurred before and during the fever. At the same time there is a gradual disappearance of the indifference noticed in the patient during the disease, so that surrounding objects begin to attract attention, and the moral feelings to be awakened.

With this improvement in the condition of the nervous system, the muscular powers are strengthened; the pulse becomes calm, regular, and open, though frequently weaker than in health; the heat of the skin, perspiration, and thirst disappear; the appetite improves, and the nights are passed in comfortable refreshing sleep.

It has been a very common opinion from early times, that in fevers, the decline and disappearance of the symptoms is accompanied or preceded by a crisis (a term implying *separation from the body*). This doctrine, however, cannot be applied to enteric fever, and for the obvious reason, that its progress and duration are so intimately dependent on the lesions peculiar to this disease, as to preclude the notion of its duration being influenced by any critical or depurative evacuations from the system. The doctrine of critical days, therefore, must be considered in reference to typhus fever without secondary affection only.

A change in the aspect of a case of typhus, favourable or otherwise, is doubtless in many cases preceded by sweating more or less profuse and persistent, sometimes by copious deposit of lithates in the urine, by spontaneous diarrhœa, some form of cutaneous eruption, or occasionally, by purulent formations in one or more parts. This theory receives apparent support from comparison with other forms of fever,—the periodic and relapsing, for example—in which there is an abrupt termination of the febrile disturbance after a profuse and apparently salutary sweat.

The law of periodicity in intermittents—the precision with which the symptoms recur and ter-

minate, as well as the crisis with which the solution
of a periodic fever is almost invariably accompanied,
are certainly among the most singular phenomena
in the whole range of practical medicine.

When we take into consideration, however, the
large proportion of cases in which the convalescence
slowly, and, as it were, imperceptibly proceeds,
without any evident critical event, we cannot resist
the conclusion that, as a general rule, the fever
leaves the body without any marked evacuation,
and that when a crisis does occur, it is to be
regarded as an exceptional, though favourable
circumstance.

Again, it has been affirmed, that the change
either to convalescence or to a fatal issue, happens
on particular days, in preference to others, and
hence the origin of critical days.

I may for a moment recal to your notice, that
the traditional days on which the termination, or
crisis, has been supposed to occur, are the third,
fifth, seventh, ninth, eleventh, fourteenth, seventeenth, and twenty-first. The non-critical days
are stated to be the intermediate days; and, as if
to give this theory the appearance of being founded
on exact observation, the fourth and the sixth have
been considered secondarily critical.

It is doubtful, however, if the doctrine of critical
days, as applied to continued fevers, has been
hitherto tested by accurate observation. In dealing
with this subject, several circumstances should be

kept in view. For example, it is an admitted
fact, that till within a recent period the types or
forms of fever were unrecognised, while the various
secondary lesions (complications), when they do
occur, materially influence or interfere with the
ordinary progress of the disease. We know, too,
that there is often much difficulty in obtaining such
a history of the case as to warrant its commencement or invasion being accurately determined, besides that the intensity with which the febrile poison
operates, is by no means uniform, being sometimes
sudden and severe, sometimes slow, and followed
by trivial effects. Again, in drawing conclusions
from a large number of cases, with the view of determining this question, there is a possibility that a
proportion of acute diseases, not strictly fevers, may
have been included, besides that the methods of
estimating the duration of the fever are seldom
alike, some, for example, fixing the termination,
not at the cessation of the symptoms or commencement of convalescence, but at the period when the
health is completely re-established.

Lastly, in enteric fever, as has just been pointed
out, the lesions with which it is associated render
its duration so uncertain, that it cannot be defined;
it may terminate, as we have seen, in the first,
second, or third week, or it may even be protracted
to fifty or sixty days, and even beyond, the duration depending mainly on the progress and extent
of the intestinal lesions.

Before proceeding to describe the morbid appearances found after death in typhus fever, I may
allude to the opinion entertained by some that the
morbid anatomy of this disease has not been subjected to the same rigorous scrutiny as that of
enteric fever. This remark may be true in respect
of the investigations of continental pathologists,
as contrasted with their successful elucidation of
the morbid anatomy of the latter disease. I
cannot, however, admit that the pathology of
typhus has been overlooked by British observers,
though their labours have been more particularly
directed to recording the circumstances by which
various epidemics have been distinguished, more
especially their prevalence in certain places, the
numbers affected, etiology, and modes of treatment, on which points much valuable information
has been collected. But it cannot be denied that
much of the obscurity and confusion that have prevailed in regard to the morbid anatomy of fevers
in times not very remote, may be ascribed to the
hesitation with which the doctrine of the identity
or non-identity of the two principal varieties of
the ordinary fevers of Britain—typhus and enteric
fever—has been received. The distinction is, even
in the present day, disputed by many, who, in de-
fiance of the plainest illustrations and the most
logical deductions, adhere to the untenable doctrine
that the two forms of fever are essentially one and
the same disease; and notwithstanding the search-

M

ing scrutiny which their pathology has of late years undergone.

In examination of the bodies of persons who have died from typhus fever, without secondary or inter-current local affection, in a considerable proportion, after the most careful search, no trace of disease in any organ or tissue can be discovered. In other instances, the only deviation from the normal state is referable to congestion of the internal organs generally, but especially of the mucous surfaces.

When secondary affections, however, have super-vened, various lesions are observed after death.

In the brain, engorgement of the sinuses and larger veins is not uncommon. Congestion of the cerebral substance, shown by an increased number of red points on the cut surface, is found in nearly half of the cases examined. It is also less firm than natural, so that on removing it from the calvarium portions of cerebral tissue adhere to its investing membranes. This diminished consistence may be limited to the external or cortical substance, or it may be found also in the central parts—the fornix and septum lucidum.

The membranes are more frequently altered than not. The pia mater often bears evidence of congestion of both large and small vessels. With this congestion, there may be a varying amount of subarachnoid serosity on the surface and at the base of the brain, or there may be more or less limpid fluid in the cavity of the ventricles.

The occasional occurrence of hæmorrhage into the cavity of the arachnoid in typhus fever has been pointed out by Dr. Jenner. He found this lesion in five out of thirty-nine cases, in which the head was opened. In every instance, the coagulum was in the form of a delicate red film, varying in thickness and colour, and situated on the convex surface of the brain. In one, it consisted of two or three delicate fibrinous films only. In two of the five, the coagulum was confined to the right side; in three, it was double—that is, existed on both hemispheres. In one of the five, it was accompanied by effusion of blood into the substance of the rectus abdominis muscle. The brain substance was apparently healthy in all, and as no aperture could be found from which the blood had escaped, the source of the hæmorrhage could not be ascertained.

If we examine the accounts of the morbid appearances in the brain in typhus in an excellent report* by the late Dr. John Reid, of the epidemic fever of Edinburgh, (1838-1839) we find, that of forty-three cases which came under his observation for examination after death, there was increased effusion, of varying amount, in 25 (or in more than a half), the most common seat being under the arachnoid; in the remaining eighteen the quantity of serum was not more than usual. In all the cases

---

* Edin. Med. and Surg. Journ., vol. 52.

except one, the vessels of the brain were well filled with blood, and in many, a greater number than usual of red points were observed on the cut surfaces of the brain. The cerebral substance was of natural consistence in all except one, and in this case the softening, which was slight, extended throughout the whole brain. Dr. Reid endeavoured to trace the connexion between the symptoms during life and the appearances after death in these 25 cases, and after a careful review of the morbid appearances, and of the symptoms exhibited during life, and also contrasting those cases in which an abnormal amount of serum was found within the cranium, with those where the usual amount only was observed, he came to the conclusion, that no distinct evidence was afforded that the serous effusion was in all, if in any, the cause of death. In one case, indeed, in which an ounce of serum was found in the ventricles, nothing but the usual mental confusion was observed during the fever. The cerebral derangement, too, was as strongly marked in those cases in which there was no increased effusion within the cranium, as in those where it was found. He also thought that when the vessels of the brain are loaded with blood, the apparent vascularity may be partly due to the increased fluidity of the blood, while in cases in which there is severe pulmonary affection the impeded circulation through the lungs may re-act upon the brain, and thus give rise to more or less cerebral congestion.

The pharynx and larynx, in the examination made at the Fever Hospital, were both found normal in seventeen out of twenty-six cases. In five, Dr. Jenner noted signs of inflammation of both larynx and pharynx, and in one of the five, there was ulceration of the larynx. The larynx alone was inflamed—that is, without co-existing lesion of the pharynx—in one instance only; but in no instance was the pharynx diseased and the larynx healthy. In not a single instance was there ulceration of the pharynx—typhus showing in this respect a marked difference from enteric fever, in which latter it occurred in one-third of the bodies examined after death.

In fatal cases of typhus, the pulmonary organs are rarely found in a normal condition. In a considerable proportion the mucous membrane of the bronchial tubes is more or less congested, indicated by dark-red injection of the membrane, with increase of the ordinary secretion, or with frothy serosity.

In the pulmonary tissue, the principal lesion is congestion, generally of the posterior portions, sometimes of one, sometimes of both lungs, and with or without diminished consistence of the affected parts. This congested state may be accompanied with consolidation, constituting non-granular lobular congestion; but the granular lobular consolidation so frequently found in enteric fever is much less common—indeed, it is rarely observed in typhus.

In other cases, again, the only change observed is

colourless serosity in some portions of lung, often in the upper lobes, the depending or inferior being more or less congested.

Pulmonary gangrene, an incidental, though somewhat rare lesion in typhus, is readily recognised by the peculiar fœtid odour and complete destruction of the portion of lung tissue involved. It would appear to be due less to pre-existing inflammation than to the action of a specific poison on the lungs. It is generally found in one or more circumscribed portions of irregular form, of a dark-brown or greenish colour, and more frequently at the summit than at the base of the lung. The surrounding pulmonary tissue has an unhealthy appearance, and when cut into is soft and lacerable, as if ready to pass into mortification. Sometimes a pulpy membrane may be seen on such portions of the sphacelated mass as have become detached, and from which a thin bloody sanies has been secreted.

In some subjects, serous fluid in varying amount is found in the pleural cavity. As pleuritic inflammation rarely occurs in typhus, the pleural membrane seldom shows evidence of it; and when it has happened, instead of the usual coating of lymph on the membrane, we find detached shreds mixed with the fluid effused in the pleural sac, the pleural surfaces sometimes adhering by a thin layer, which is soft and easily detached.

In the examinations made at the Fever Hospital

of thirty-five cases, the lungs were free from disease in two or three cases only. In the report of Dr. Reid already alluded to, he found that in every case examined there was some form of pulmonary lesion. Thus in ten, there was increased effusion of mucus into the bronchial tubes; in fifteen, congestion in the lower portion of the lung; in one, there was recent pneumonia of the right lung; and in thirteen, the posterior and central portions of both lungs were gorged with a frothy serum, some portions being so dense as not to crepitate when cut, though they did not present any granular appearance. These appearances resembled those observed after section of the vagi nerves, and were in all probability principally due to the embarrassed respiration consequent on derangement of the central parts of the brain, and which generally occur a short time before death. Or they might have arisen from the enfeebled action of the right side of the heart, for when its action is weak, the bronchiæ somewhat obstructed, and the blood more fluid than usual, the right side of the heart becomes unable to propel the blood through the lungs, and the blood consequently accumulating in the more depending parts, the same results follow as when the respiratory movements are deranged.

There is in general no alteration in the heart, except in the more severe and malignant forms of typhus, in which its muscular substance is found soft

and lacerable, and evidently partaking of the diminished consistence of the solids in general. The
lining membrane is frequently stained of a dark or
purple-red colour. The blood found in its cavities,
as well as in the vessels generally, is either
fluid or in soft masses, showing its defibrinated
condition.

In regard to the abdominal organs, the great
pathological difference between typhus and enteric
fever consisting, as has been repeatedly shown, in
the invariable absence in the one, and the equally
invariable presence in the other, of distinctive
lesions, we find in typhus, with the exception of
occasional softening of portions of the mucous
membrane of the stomach, the whole tract of the
alimentary canal in a normal state. The mesenteric glands are also free from disease. In the
large intestines, with the exception of the mucous
membrane of the cæcum and colon being sometimes of a deep-red colour, from increased vascularity, there is no deviation from their ordinary
natural condition.

The spleen is normal in about half the cases
examined after death. In the other half, it is
simply softened, without increase of volume or
size, a small proportion only exhibiting both softening and enlargement. From the dissections performed at the Fever Hospital by Dr. Jenner, it appears that splenic lesions are found both in typhus
and enteric fever, but the spleen is larger in the

latter than in the former; in typhus, it is much less enlarged in those above the age of forty than in those under that age. Splenic softening is also more frequent in typhus after the age of fifty than before, and this condition is more often found when death takes place before, than after the fourteenth day of the disease. It seems, too, that in both forms, splenic softening is more frequently observed when death takes place at an early than at a late period of either disease.

The liver does not undergo any special change that can be ascribed to the agency of the febrile poison. It may partake of the general softness of the solids, or when cut into, may be gorged with thick dark blood; but whether this congestion existed during life, or is due to changes occurring in the act of dissolution, is yet undetermined.

This short summary of the morbid anatomy of typhus leads to the inference, that there are no special lesions that can be said to be peculiar to this form of fever. The most common, and perhaps the most constant changes, seem to be due to congestion of organs, more especially of their mucous surfaces; but that even these changes are not an essential or invariable element of typhus is proved by the fact, that in a large number of bodies, after the most careful examination, no trace of disease in any organ or tissue can be discovered. Pathologists have therefore turned their attention to the condition of the fluids, with the view of connecting the phenomena of fever with such deviations from their normal or healthy condition as the present more advanced state of chemical science has elucidated.

The blood, which from remote ages has been supposed to undergo marked changes in its physical properties in fevers, has been in more recent times subjected to chemical analysis; and these investigations show that several of its constituent principles become altered, either primarily or secondarily.

We know from the researches of physiologists that healthy blood contains, besides water, four essential constituents—corpuscles, fibrine, albumen, and certain saline matters. The fibrine, albumen, salts, and water, constitute the liquor sanguinis— a transparent colourless fluid, in which the corpuscles, red and white, are suspended and carried along in the living vessels.* When blood is drawn from the body, as in ordinary venesection, and allowed to remain at rest, it separates into crassa-

---

* In healthy blood—
  Fibrine,
  Albumen,
  Salts,
  Water, } constitute the liquor sanguinis.

In coagulated blood—
  Fibrine,
  Corpuscles, } constitute the crassamentum.

  Albumen,
  Salts,
  Water, } constitute the serum.

mentum and serum; the crassamentum consisting of fibrine, in which the red and white corpuscles are blended together with a small quantity of serous fluid, while the serum differs from the liquor sanguinis in one respect only—that it does not contain fibrine; it contains, however, albumen, which gives it the property of coagulating by heat; and if a degree of heat sufficient to decompose the animal matter be applied, the residuum becomes converted into earthy and alkaline salts.

In regard to the proportional or mean amount of these constituents, Becquerel and Rodier found that in 1000 parts of healthy blood there are, of

Water	791·1
Fibrine	2·2
Corpuscles	127·2
Albumen	70·5
Fatty matters	1·6
Extractive matters, salts, and loss	7·4
	1000·0

This outline of the constituents of healthy blood may serve for comparison with the changes which it has been observed to undergo in fevers. But it is to be understood that the few observations I shall make have reference to the blood in both forms of fever (enteric and typhus), the difference between the constituents of the blood in the one or the other being trivial.

The blood in continued fevers contains a smaller amount of fibrine than healthy blood, besides being often deficient in albumen also. Hence, when bloodletting has been resorted to in the treatment, and more especially when performed in the advanced stage, it coagulates imperfectly, the crassamentum being soft and diffluent, and rarely exhibiting the buffy coat. It is likewise more liable to putrefy than healthy blood, or than blood abstracted in other diseases. This deficiency in the fibrine of the blood in fevers, when uncomplicated, tends to distinguish it from that in acute inflammations, in which the amount of fibrine is increased in proportion to the intensity of the local disease.

This gradual defibrination of the blood in fevers cannot, as has been supposed, be due to the antiphlogistic measures practised in their treatment, since in other acute diseases, bloodletting (seldom, now-a-days, employed as a remedy in fevers) and other depletives do not appear to be followed by diminution—nay, rather by augmentation—of the fibrine. It has moreover been ascertained, that if a local inflammation spring up in the progress of fever, the amount of fibrine becomes increased, though it is still less than in ordinary inflammation, the febrile condition apparently influencing the quantitative amount of this constituent, when an intercurrent inflammation has arisen. Besides, as soon as the period of convalescence arrives, the fibrine

begins to increase, and before the prescribed nourishment could be so elaborated as to contribute to this
change, and which appears to proceed with advancing convalescence.

With this diminution of fibrine, it has been
shown by the experiments of Andral and Gavarret,
that in fevers, a considerable increase in the corpuscles takes place, forming another point of contrast
with the blood in primary inflammations, in which
there is a slow but certain diminution in the corpuscles.

It may therefore be concluded, first, that in continued fevers, the fibrine gradually decreases in
proportion to the duration of the fever, never
exceeding in amount the normal standard, but
being much more frequently below it; and, second,
that the corpuscles become increased in amount.

It appears, too, that the residue of the serum
and solid constituents generally, are little if at all
diminished in amount, notwithstanding the statements made by some writers, that according to the
duration and intensity of the fever, the salts of the
blood, especially chloride of sodium, become de-
creased, a notable increase of water at the same
time taking place.

In regard to the supposed origin of fevers in
morbid changes in the blood, I may observe, that
when we reflect that the blood, which was aptly
termed by Bordeu fluid flesh, has the same proximate principles as the solids of the body, that it is

organized, and apparently endowed with vitality, the doctrine, that it is liable to become diseased, rests on a more certain foundation than mere vague hypothesis. The causes by whose agency the phenomena of fever are induced, have hitherto, notwithstanding many ingenious explanations, been veiled in mystery. They doubtless, however, produce sensible effects on the blood, but whether the fever poison acts primarily on this fluid, producing changes in its sensible properties, or whether, as some are inclined to believe, the changes are secondary, is a matter difficult to determine, and of no practical importance. Indeed, we have no data to guide us in this inquiry, except the statements of Dr. Stevens, who affirms that he and other practitioners in the West Indies observed, that in persons exposed to the fever-producing malaria, the blood underwent marked changes, even before the symptoms became developed.

I am inclined to believe that the different forms of fever arise from distinct poisons, however induced; that the true typhus poison acts on the blood alone, that enteric fever is the result of a poison *sui generis*, which has a compound action on the blood as well as on the solids, and by which the special lesions with which this type of fever is invariably associated are induced; and that it is more than probable that the relapsing fever, to be hereafter described, has also its origin in a distinct poison, though from its comparative infrequency, little or nothing, as yet, has been satisfactorily made out as to its etiology. And let me observe, that no department of the pathology of fevers is more important, and more worthy of further elucidation, than their connexion with morbid states of the blood.

The importance of forming a correct diagnosis between the two forms of fever (enteric and typhus) cannot be over-estimated, the one differing from the other not only in many prominent circumstances, but likewise, as we shall afterwards show, in their treatment.

In order to arrive at a correct conclusion as to their differential diagnosis, it will be necessary to institute a short comparison of the individual symptoms of each form.

1. It has been pointed out, that in typhus the accession of the symptoms is more sudden than in enteric fever, in which it is generally gradual, slow, and insidious. The prostration, or muscular weakness, is also from the first more marked in typhus.

2. The heat of skin is more constant and more marked in the early stages of typhus; during the exacerbations, it is often pungent and accompanied by flushing. In enteric fever, the heat of skin is moderate, or it may be entirely absent.

3. The cutaneous rash is quite dissimilar in the two diseases. In typhus, it is dark and mottled,

often extensively diffused over the skin, unchanged by pressure, and persists throughout the disease. In enteric fever, the rose spots are bright, isolated, slightly raised, fade or entirely disappear on pressure, and are found chiefly on the chest and abdomen. They last for a few days, and are succeeded by a fresh crop, which goes through a similar course until the termination of the fever.

4. The brain symptoms, especially delirium, intellectual dulness, and stupor, are more strongly and early developed in typhus. In enteric fever, the cerebral symptoms appear later—usually not until after the first week—and slowly and gradually increase in severity.

5. In typhus the bowels are generally confined, so as to require the occasional administration of aperients. There is seldom diarrhœa, and if the bowels act more frequently than usual, the evacuations are not watery. There is, besides, no gurgling on pressing the region of the cæcum, and rarely meteorism. In enteric fever, the evacuations, on the other hand, are generally frequent and watery, and accompanied with griping. The belly is tense, generally inflated, and gurgling noise may often be detected by pressure over the right iliac fossa.

6. Epistaxis is rare in typhus, not uncommon in enteric fever.

7. Intestinal hæmorrhage is very rare, almost unknown, in typhus; it is frequent in enteric fever.

8. Emaciation is much less marked in typhus

than in enteric fever; in the latter, when protracted, it is often extreme.

9. The anatomical characters (or lesions) are
distinctive. In typhus, Peyer's patches, and corresponding mesenteric glands, are in their natural
condition. The vessels and sinuses of the brain
are more generally gorged with dark blood,
and the lungs exhibit more frequently evidence
of congestion or other changes, especially of the
posterior and inferior portions, than in enteric fever.

In enteric fever the alterations in Peyer's patches
and corresponding mesenteric glands form the
great distinguishing characteristics of the disease.
The spleen is more often enlarged and softened.
Ulceration of the pharynx and œsophagus, and
softening of the mucous membrane of the stomach,
are more common than in typhus.

10. The duration of the two forms may also
serve as diagnostic marks. In typhus, it is
generally less protracted, terminating either in
recovery or death within the second week, much
more frequently than in enteric fever. I have
stated that the average duration of typhus does
not exceed fourteen or fifteen days, or, if prolonged beyond that period, it is generally owing to
some secondary affection or complication. In enteric fever the average duration is longer,—from
twenty to thirty days, extending, as has been shown,
in some instances, to forty, fifty, or sixty days, or
even more.

N

It is a matter of common observation that in
children there is more difficulty in determining
the diagnosis between either of the forms of fever
and various acute diseases, than in adults. Let
me mention a few of the more prominent illustrations of this remark.

Disorders of the gastric system attended with
fever are very common, especially among the badly
nourished children of the poor, and often simulate
enteric fever; or, perhaps, as not unfrequently happens, the latter is mistaken, and treated as an acute
gastric affection by irritating purgatives. If, however, the symptoms that are peculiar to enteric
fever be kept in view, the distinction is less
difficult. For example, the invasion is not sudden,
nor can the symptoms be traced to dietetic misma-
nagement, the prostration is more marked, the child
is at first restless and wakeful, afterwards som-
nolent and delirious in the night, the stools are
frequent and watery, the belly tender and inflated,
and the splenic region dull. But above all, the
characteristic rose spots, if present, and the prolonged duration of the pathognomonic symptoms,
are sufficient to indicate that the child is suffering
from enteric fever.

Enteric fever resembles in some respects infantile gastric remittent. It may be distinguished
however by the irregularity of the paroxysms in
the latter, by the absence of rose spots, by the
large, often solid evacuations, requiring active

aperients for their dislodgement, by the occasional admixture of undigested aliment in the secretions expelled from the bowels, and by decided improvement resulting from a course of treatment that would certainly aggravate, if it did not hopelessly augment, enteric fever.

It is right however that I should mention, that many physicians of the present day disbelieve in the existence of gastric remittent fever, without co-existing affection of Peyer's patches, which they state to be a constant element in this common infantile disorder. I cannot adopt this conclusion—at least I am unwilling to do so—until there be better evidence by accumulated observations of the appearances in the intestine, when this form of gastric fever proves fatal, which, I may observe, rarely happens—a circumstance to be regarded as another distinctive character, and showing clearly, to my mind, the non-identity of the two diseases.

Another disease of childhood, which it may be difficult to distinguish from enteric fever, is meningitis, and more especially that form which occurs in strumous children — tubercular meningitis. In this affection, the symptoms from the very commencement indicate severe cerebral disease— violent pain in the head, flushing, aversion to light, and, as the disease advances, drowsiness, delirium, dilated pupils, and finally deep coma, sometimes alternating with convulsions. There may be sickness, and perhaps vomiting, but the bowels are generally

confined.   The rose spots, too, so characteristic
of enteric fever, are absent.

If we compare these symptoms with those of
enteric fever, it will be apparent that the cerebral
symptoms are much less prominent in the latter,
at all events in the early period, that there is
seldom sickness or vomiting, almost invariably
diarrhœa, meteorism, dry, fissured tongue, often
rose spots, and when the fever is protracted, exudation of sordes on the lips and teeth, progressive
and often extreme emaciation, sloughing of parts
subjected to pressure, and death by gradual exhaustion, rarely, however, preceded by convulsions.
The duration of the two diseases is also dissimilar.   Meningitis terminates either in recovery
or death within the third week, while enteric fever
may be protracted to the thirtieth or fortieth day,
or even much beyond.

There is less difficulty in distinguishing infantile typhus.   The absence of gastric symptoms,
the less pronounced cerebral affection, the less pungent heat of skin, the morbillous eruption if
present, the propagation of the disease by contagion, and its comparatively short and definite
duration, are generally sufficient to distinguish
typhus fever from other acute disorders of infancy
and childhood.

RELAPSING OR RECURRENT FEVER.

Historical sketch.—Nature, symptoms and progress—anatomical characters.

THIS peculiar form of fever—not of frequent occurrence, but appearing now and then epidemically—derives its name from a relapse or recurrence of the symptoms after a definite period of convalescence, so that there are apparently two distinct fevers blended in one disease.

In the introductory observations to these lectures I stated, that relapsing fever had been recognised by Dr. Rutty more than a century ago (in 1739), and alluded to by him in his History of the Diseases of Dublin. "In July, August, September, and October (writes Dr. Rutty), a fever prevailed which terminated in four, for the most part in five or six days, sometimes in nine, and commonly in a critical sweat: it was far from being mortal. I was assured of seventy of the poorer sort at the same time in this fever, abandoned to the use of whey and God's good providence, who all recovered. The crisis, however, was very imperfect, for they were subject to relapses, even sometimes to the third time." He describes the same fever as occurring also in 1740, 1745, 1764, and 1765; remarking that in the fever of 1765 the bowels were in some instances remarkably affected.*

In the records of several Irish epidemics we find frequent allusion made to relapses after apparent convalescence, but from want of more precise details, we can only infer that sporadic cases of relapsing fever had been observed during the epidemic visitations of the time. This seems to have been the case more particularly in the epidemic fever of 1817, 1818, and 1820, which, from the crisis almost invariably taking place on the fifth day, was popularly termed the "five days' fever." The disease almost invariably terminated by copious sweating.

From the reports of Drs. Barker and Cheyne of the fever that prevailed in the same years, relapses seem to have been very common, for at Cork, the numbers who relapsed are computed at 2000, and at Waterford, it was estimated that a fifth or sixth of those who were attacked with fever relapsed.

In describing the fever of 1826, Dr. O'Brien mentions, that it was distinguished by its short periods, terminating in three, five, seven, or nine days; but the second of these periods was the most frequent. The individual, previously in perfect health, was seized with sickness at stomach,

---

* A Chronological History of the Weather and Seasons, and of the Prevailing Diseases of Dublin. By JOHN RUTTY, M.D. London, 1770. pp. 75 and 76.

headache, pain in the small of the back, and chilliness. On the approach of the evening, all these symptoms increased, and the febrile paroxysm was formed; the chilliness increased to a rigor, the nausea to vomiting, which harassed the patient during the first three or four days, in the form of empty straining, and frequently continued throughout its whole course. On the evening of the fifth or seventh day, the exacerbatio critica commenced, which, mostly with the intervention of a rigor, but very frequently without this symptom, terminated in a profuse perspiration, which continued through the night, so that on the following morning the crisis was complete, and the patient generally became convalescent.

These particulars correspond with the epidemic relapsing fever of Edinburgh, except that, in the latter, the abrupt termination happened in the majority of the cases on the seventh day.

We possess excellent histories of this form of fever, as it has been observed in different parts of Scotland, at various periods within the last forty years. It constituted a large proportion of the Edinburgh epidemic (1817-20) described by Dr. Welsh,* and seems to have been even more common in Glasgow a quarter of a century afterwards; for we find in the published records of the

---

* Practical Treatise on the Efficacy of Bloodletting in the Epidemic Fever of Edinburgh.

Glasgow Infirmary that in ten years—from 1843 to 1853—no fewer than 7804 cases of this relapsing fever were admitted into this hospital; and when we reflect that this form constituted a proportion only of the fever cases, we have some idea of the prevalence of epidemic fever in Glasgow in those years.

I had the opportunity of witnessing this fever when it appeared in Edinburgh in 1817-20, in the description of which, after detailing its prominent features, Dr. Welsh states, that of 743 tabulated cases, 183 relapsed—being in the proportion of one in five; but nothing is mentioned that would lead to the inference that these relapsing cases were to be regarded as a distinct or peculiar type of fever; indeed, this monograph appears to have been written rather with a view of showing the supposed advantages of indiscriminate and unnecessary waste of blood in its treatment, than of directing attention to the peculiar features of a form of fever differing in many respects from previous epidemic visitations.

Dr. Wardell,* in an excellent account of the Scotch epidemic fever of 1843-44, mentions that of 980 cases, 603 had one or more relapses; 67 had two relapses; nine had three; and one had four; and there were one or two other instances in which the patient had no fewer than five separate and

---

* Medical Gazette, vols. xxxvii. and xl.

distinct attacks. From this account it appears, that not less than two-thirds relapsed before leaving the hospital; and, when it is considered that many might have a return after their discharge, it would be no exaggeration to assume that three-fourths relapsed.

We find too from Dr. Cormack's work,* that anti-periodic remedies, administered with the view of preventing relapse, were successful in a limited number of cases only. It appears moreover, from the cases observed at the same period and reported by Dr. Craigie, that 110 out of 182 relapsed, or about 60½ per cent.

In and around London this relapsing fever has been less frequently witnessed. By referring to the table of admissions into the London Fever Hospital, we find that of the numbers of all forms received in the ten years between 1848 and 1857, there were only 441 cases, and since 1854 there has not been a single example of it.

From these details it may be inferred, that this somewhat singular disease has never constituted the character of an entire epidemic, but only of a proportion of the cases. It is therefore extremely probable that it originates from a poison *sui generis*, and that it is contagious does not admit of doubt. Moreover, if we examine closely this fever as it

---

* On the Epidemic Fever at present prevailing in Edinburgh and other Towns.

has prevailed at different times, though there may have been a great resemblance in the general features of the disease, a marked difference in epidemic constitution is apparent; so that, as has been pointed out by Dr. Wardell, the relapsing fever of 1817-20 differed materially from that of 1843-44, in so far that the former was much more acute in form than the latter, in which, too, the relapses were far more frequent, and the cases accompanied with yellowness of the skin more numerous. This shows that in relapsing, as in other forms of fever, differences in type, more or less marked, take place, and require to be carefully observed when an epidemic first breaks out.

In regard to the nature of this fever, Dr. Wardell states (*op. cit.*), " there were undoubtedly some considerations which led to the supposition that the epidemic relapsing fever bore certain resemblance to the *suette*, or sweating fever of Normandy. In a few instances, though these were of rare occurrence, the epidermoid tissue was raised into vesicular eminences, varying from the size of a millet seed to the section of a small pea, these vesicles containing a transparent fluid, and quite unattended with any areolar blush. On the third day they became shrivelled and opaque, and desquamated in thin furfuraceous scales. From the occasional presence of these bullæ, with other more logical characters, some degree of similarity certainly was manifest between it and the *suette*. There were physicians

who endeavoured to show its near alliance to the
yellow fever of the West Indies—indeed gave it
as their opinion that, in some respects, there was a
positive identity between the two, only that the
epidemic prevalent in this country had become
greatly modified by climate and other circumstances
calculated to alter its general features. When we
take into consideration the unusual number of yellow cases, together with two or three cases of less
important correspondent symptoms, we are compelled to admit that the assertion is not wholly
unfounded. No trace of its importation into Scotland, however, could be found, which has generally
been the case where yellow fever has been communicated from one country to another."

I have always regarded relapsing fever as a form
intermediate between the continued and periodic,
but having a more close analogy to the latter. I
have been led to this view by considering the suddenness of the invasion, the abrupt termination of
the symptoms, after a definite period, by copious
and apparently critical sweat, the interruption of
the convalescence by a similar though shorter paroxysm, or it may be paroxysms, of nearly certain
duration, and a final somewhat sudden cessation of
the disease, generally after critical sweating. Even
in the more severe cases in which there is gastric
disturbance, with jaundice, and occasionally cerebral
symptoms, the resemblance to the malignant or
pernicious periodic fevers, more particularly those

included under the bilious remittents of tropical climates, is striking.

As to the characteristic symptoms of relapsing fever, the invasion is sudden. The patient, previously in good health, without warning, is seized with a feeling of indisposition, complaining of chilliness or shivering, acute headache, languor and lassitude, severe muscular aching, and arthritic pains. The appetite fails, the skin becomes hot and dry, the tongue white, and the desire for fluids constant. Towards evening, the symptoms are aggravated; the night is passed either in restless agitation or with snatches of unrefreshing sleep. Occasionally, the heat of skin is relieved by irregular sweating, but still the other symptoms suffer no diminution. Vomiting of bilious fluid, often accompanied with pain at the epigastrium, is an early and nearly constant symptom. It may occur in the first or primary fever only, or it may come on in the relapse also. As the disease progresses, the patient becomes more prostrate and disinclined for bodily or mental exertion, the pulse more rapid and tense, the tongue more thickly coated, the urine scanty, the bowels constipated, the muscular and arthritic pains more acute; and the nights are passed in restlessness and wakefulness, unless the nervous system be calmed by opiates.

About the third day a marked remission of the symptoms is often observed; but whether there be a remission or not, at a period varying from three to seven days—more commonly on the fifth day—a copious general sweat breaks out, and almost immediately afterwards the whole phenomena of the fever suddenly vanish, leaving the patient unexpectedly free from the painful symptoms with which, a few hours previously, he had been harassed. Dr. Cormack, who watched closely the phases of this singular disease, tells us that the change for the better was often so sudden and complete, that the patient who was one day moaning and groaning in pain, was on the next at his ease and cheerful, complaining only of hunger and weakness.

This apparent convalescence, however, is not of long duration, for when the patient and his medical attendant reasonably conclude, that from the favourable change that has occurred the fever is at an end, and that time only is required for complete restoration to health, a sudden and unlooked-for recurrence of the previous symptoms takes place. This relapse happens at some period between the twelfth and the twentieth day (from the beginning of the disease,) or on or about the seventh after the crisis, and without apparent cause or indiscretion on the part of the patient. The relapse is indicated by the same symptoms as the primary fever—rigors, headache, muscular aching, hot skin, thirst, quickened pulse (the rapidity being often disproportionate to the other symptoms), coated tongue, and loss of appetite.

After a few days—two, three, four, or five—this second attack suddenly ceases after a profuse sweat, and the patient becomes a second time convalescent. The return to health is comparatively rapid and complete in the young and vigorous, but in the aged, and especially in those who have been previously in indifferent health, strength is more slowly regained.

Nor does the mildness or severity of the relapse appear to be influenced by the previous attack; for it has been observed, that the symptoms of the second are in some cases more mild, in others more severe, than those of the primary fever. In some instances, for example, in which the first attack was by no means severe, the second has been characterized by delirium, deep jaundice, violent purging, and other grave symptoms. Such cases are, however, exceptional.

Sometimes again, a second but mild relapse takes place, generally about the twenty-first day, and I have already alluded to the circumstance, that patients sometimes have suffered three, four, and even five separate and distinct attacks. Such frequent relapsings, however, have been seldom noticed in the epidemics in England.

Other anomalies are in some instances observed. Thus, the symptoms of the relapse, instead of appearing suddenly, come on gradually and insidiously; or instead of the ordinary well-marked progress of the symptoms, there may be only slight acceleration of pulse, and a little increased heat of skin, to mark its occurrence. Occasionally, in place of the abrupt termination of the attack by sweating, the crisis has apparently been connected with some other evacuation, such as diarrhœa, hæmorrhage from the nose, or the menstrual discharge. In some cases, on the other hand, in which the ordinary symptoms of the first attack have been well marked, there has been no relapse, nor anything approaching to a recurrence.

There appears, too, in pregnant women, to be a greater tendency in relapsing fever than in other acute diseases to abortion or premature delivery. So invariably, indeed, according to Dr. Wardell's experience, did this happen, that throughout the whole duration of the Edinburgh epidemic,—a period extending over at least fourteen or fifteen months,—he never observed even a solitary instance in which the impregnated uterus did not expel its contents; and the statements of others, whose opportunities of observing this fever were equally ample, confirm the statement. The same tendency to abortion was observed in the patients received into the London Fever Hospital.

But it should be kept in mind that the relapsing fever sometimes assumes a more severe character, the aspect of the symptoms from the commencement indicating a much more serious disease. The rigors are violent, the heat of skin is intense, the heart's action depressed, indicated by the softness and compressibility of the pulse, the patient complains of extreme prostration and feeling of exhaustion or sinking, there is often incessant vomiting of bilious fluid, accompanied with a more or less deep jaundiced appearance of the skin, though the evacuations from the bowels exhibit no deficiency of bilious admixture, the urine however being generally loaded with bile.

In some cases, sudden collapse takes place—the pulse becomes rapid and feeble, the skin universally cold, more especially the hands and feet, the face livid, partial or complete unconsciousness succeeds, the sphincters become relaxed, and after a few hours, death takes place.

The diagnosis of relapsing fever may be given in a few words.

It differs from other forms of fever—1, by its sudden invasion; 2, by the short duration of the primary fever, and its termination by an evident crisis; 3, by the almost uniform occurrence of a relapse—occasionally a second or third; 4, by the unusual number of cases with more or less jaundice or yellow colour of the skin, accompanied often with gastro-enteric and gastro-splenic symptoms; and 5, by the absence of characteristic rash.

The small mortality, or death-rate, of relapsing fever, being about one in twenty-five, or under 3·9 per cent, shows its comparative mildness,

This singular form of fever, if uncomplicated,

seldom proves fatal. In examination of the fatal cases, no special lesion, so invariably present as to indicate the anatomical character of the disease, has been discovered.

The blood has in some cases been found throughout the body in a fluid state, indicating a decrease in the normal amount of fibrin. I am not aware that it has been subjected to further chemical analysis.

The brain, with the exception of a moderate amount of subarachnoid serosity, and perhaps an increased quantity in the ventricles, shows no remarkable deviation from its natural state.

The heart and lungs exhibit no evidence of disease.

The liver has been found enlarged from congestion, and the gall-bladder more than usually distended with bile; and a point worthy of note is, that there is no apparent obstruction to the free escape of the bile through the ducts, even in cases in which the jaundice had been well marked.

No disease in any portion of the alimentary canal is discoverable.

The spleen exhibits the most marked and constant lesion, more especially as to size or volume. In the Edinburgh epidemic (1843-4), Dr. Wardell noted several cases in which this organ was three or four times larger than natural: in one it weighed twenty ounces. This splenic enlargement has been met with by other observers: thus, in a fatal case examined at the London Fever Hospital, the spleen weighed thirty-eight ounces. It therefore appears that this organ is occasionally larger in relapsing than in either typhus or enteric fever.

Urea has been found in the blood. In a fatal case recorded by Dr. Wardell, in which the patient fell into a state of stupor twenty-four hours before death, crystals of nitrate of urea were discovered in considerable abundance in the blood abstracted by cupping, for the relief of the cerebral symptoms.

We may therefore conclude, that with the exception of splenic enlargement,—a lesion common to the other forms of fever, continued and periodic,—there is no special lesion found after death in relapsing fever. If structural changes be discovered, they are to be regarded as accidental, and due to some secondary or intercurrent affection.

FEBRICULA.

This term has been applied to a fever of short duration, and includes the ephemera,—a form of fever which generally lasts twenty-four hours, seldom more than thirty-six, though instances have occurred in which the symptoms have persisted for two or three days.

The febricula, properly so called, comes on somewhat suddenly by chills, sometimes shivering, followed by irregular paroxysms of heat of skin, loss of appetite, thirst, quick full pulse, white tongue, loaded urine, sometimes acute headache, very rarely delirium. These symptoms, more severe in some cases than in others, continue for a few days—seldom more than a week—then gradually subside, often after a moderate diaphoresis or copious lithic deposit in the urine, and the patient in a few days regains his ordinary health.

It has been supposed to arise from violent exercise, great bodily fatigue, errors in diet, or sudden mental excitement.

The ephemeral fever observed in hot climates, though of short duration, is very violent, and may even terminate fatally, by congestion of some important organ, more frequently of the brain, liver, or spleen. It is very rare in temperate latitudes. I have never seen it, but it does not appear to exhibit any peculiarity, if its limited duration be excepted; the symptoms, in fact, are those of an acute febrile disease, with abrupt termination.

But in regard to febricula as a type of fever, I must observe that though I have frequently met with cases of a short, mild disease, in which there is no special lesion or complication, at least so far as indicated by symptoms, I am not inclined to encumber our nosology of fevers with a doubtful novelty. It is a very convenient mode of classifying mild cases of either typhus, enteric, or of relapsing fever, in which the second fever or relapse does not occur. Indeed, I do not know a single pathognomonic feature of what is called febricula, except that it runs its course in a few days, and is always a very mild disease. But we have seen that ordinary typhus, as well as enteric fever, is often mild and of short duration, scarcely requiring treatment; and surely it does not require that a new nosological term should be introduced merely to accommodate such cases.

Febricula is said not to be contagious; but I apprehend that, if under this name are included cases of mild typhus, or of relapsing fever without the relapse, no one can predict with certainty that the disease will not spread.

Again, if uncomplicated, it is never fatal; but we have pointed out that secondary lesions arise in the progress of both mild typhus and enteric fever, and therefore the practitioner should be prepared for some intercurrent local inflammation springing up in cases of so-called febricula, (the more common being some form of pulmonary affection), by which its progress may be protracted.

MORTALITY OF CONTINUED FEVERS.

Mortality of all forms—tables of mortality of each form—mortality in different years— mortality of enteric fever—of typhus fever— of relapsing fever—influence of age on the mortality of fever— general deductions.

THE mortality of fevers is an exceedingly interesting subject of inquiry, and although we possess statistical data in abundance to enable us to form correct conclusions as to the mortality of continued fevers, as a class, the death-rate of the individual forms has not been satisfactorily deduced, in consequence of the several types having till very recently been considered one and the same disease. But as each form comes to be more generally recognised, we shall be enabled not only to estimate its mortality with precision, but obtain much valuable information on many points which are as yet somewhat involved in obscurity.

Continued fevers have caused at times an incredible sacrifice of human life. You may remember that I stated in my first lecture, that according to reports returned to the Government of the famine fever which raged in Ireland in 1817-19, at least 65,000 individuals were destroyed in less than two years.

The mortality varies according to certain circumstances; for example, it differs not only according to the form or type of the disease, but the death-rate of the same form varies in different years as well as in different places. Age, or period of life, too, has an important influence on the mortality. These propositions will be illustrated as we proceed.

Perhaps the most reliable information on the death-rate of fevers is the experience of the different hospitals; and I shall not, I hope, be accused of partiality if I state that the experience afforded by the records of the London Fever Hospital is exceedingly valuable in a statistical point of view, not more from the extensive field of inquiry, than from the accuracy with which the cases have been recorded, and the care with which the several forms of fever have, for the last fifteen years, been distinguished.

The mortality tables subjoined have been constructed on the experience of the last ten years only, and from the data thus afforded are drawn the conclusions which I shall briefly lay before you. And I am bound to acknowledge, that for the statistical facts I am indebted to the recently published paper of my colleague, Dr. Murchison, who, availing himself of the ample opportunities the Fever Hospital afforded, has produced a most valuable monograph on the mortality of the different forms of fever, and on the causes which apparently influence their prevalence. (*Med. Chir. Trans.* vol. xli.)

The subjoined table shows the mortality of all forms of continued fevers.

TABLE V.

Mortality of the Cases of Continued Fever of all forms admitted into the London Fever Hospital in Ten Years, 1848-57.

Years.	No. of Cases.	Deaths.	Mortality per Cent.
1848	707	148	20·93
1849	401	65	16·21
1850	361	50	13·85
1851	614	43	7·00
1852	561	50	8·91
1853	787	149	18·93
1854	714	112	15·68
1855	622	113	18·16
1856	1300	230	17·69
1857	561	99	17·64
Total	6628	1059	15·98
Deducting cases fatal within 24 hours after admission	6567	998	15·19
Deducting cases fatal within 48 hours......	6482	913	14·07

From this table it appears, that of 6628 patients admitted, comprising all the forms of fever described, the total number of deaths was 1059, being in the ratio of nearly 16 per cent., or 1 in 6¼. If from this number, however, be subtracted 61 cases, which proved fatal within twenty-four hours, the admissions are reduced to 6567, and the deaths to 998, or in the ratio of 15·19 per cent. If the cases which died within forty-eight hours be deducted, the admissions will stand 6482, and the deaths 913, or in the ratio of 14·07 per cent., or less than 1 in 7.

This rate of mortality, deduced from the experience of the Fever Hospital, including not only every form of fever, but every variety of complicated cases, is probably somewhat higher than the death-rate of a like number treated under more favourable circumstances. But it must be taken into account, as stated in almost every annual report of the Hospital, that patients are received too often in the most hopeless condition, not only in very advanced periods of life, but under circumstances precluding all hope of recovery. This statement is proved by the very large proportion of patients who do not survive forty-eight hours after admission, as well as by the numbers received in a moribund state.

In regard to the mortality in different years, the table shows considerable variation. Thus, in one year (1851), the total death-rate was 7 per cent. only; in the following year it was nearly 9 per cent.; while in 1848 it exceeded 20 per cent.: and if the mortality of one form be taken, it appears that in two years—viz., in 1849 and 1857—the death-rate of typhus cases exceeded 25 per cent.

In the paper alluded to, Dr. Murchison has made a valuable addition to our own statistics, by giving the mortality from fever of all forms, collated from the records of eleven other hospitals during the last eighteen years. Of these, one is in London, four in the English provinces, three in different large towns in Scotland, two in Ireland, (Dublin and Cork), and one in Stockholm.

From these data, we find that in St. George's Hospital (London), the average mortality from all forms of fever was 11·3 per cent., or 1 death in 8⅓; in Nottingham General Hospital, 12·78 per cent., or 1 in 7⅚; in Birmingham Queen's Hospital, 14·08 per cent., or 1 in 7$\frac{1}{10}$; in Bristol Royal Infirmary, 9·47 per cent., or 1 in 10; in the Edinburgh Royal Infirmary, 11·61 per cent., or 1 in 8⅔; in the Aberdeen Royal Infirmary, 9·14 per cent., or 1 in 11; in the Glasgow Royal Infirmary, 11·28 per cent., or 1 in 8$\frac{6}{7}$; in Dublin, Cork-street Fever Hospital, 7·46 per cent., or 1 in 13⅓; Cork Fever Hospital, 4·3 per cent., or 1 in 23¼; in the Stockholm Seraphim Hospital, 10·6 per cent., or 1 in 9$\frac{1}{7}$. On this comparative table he remarks, that the rate of mortality from fever during a series of years differs but little in the various hospitals of England and Scotland, being about 1 in 8; in some rather more; in others rather less. In the Aberdeen Royal Infirmary, however, the mortality from 8783 cases during eighteen years has been only 1 in 11. This was due to the small mortality of four years (1842-45). Taking the cases only for the last ten years, the mortality in Aberdeen, as elsewhere, was 1 in 8½.

Again, in every instance, it will be seen that the mortality has varied greatly from year to year. In Aberdeen, it was under 4 per cent. one year, and in another, nearly 18 per cent. At Nottingham, it was one year 30 per cent.; in another less than 7 per cent. The mortality at Stockholm appears to be much the same as in England, perhaps rather less.

To these results, the Irish hospitals present a marked antithesis. Out of 150,939 cases of fever admitted into the Dublin Fever Hospital, since the year 1817, only 10,632, or less than 1 in 14 died, and during the last eighteen years, the mortality has been only 1 in 13⅘. Again, in the Cork Fever Hospital, the mortality has been much less. Since the year 1817, out of 82,293 patients, only 3222, or 1 in 25½ have died; and during the eighteen years contained in the table, the mortality has only been 4½ per cent., or 1 in 23¼. Moreover, the rate of mortality has varied much less in different years than in England and Scotland. Thus, in Dublin, in no year during the last forty has it reached 10 per cent.; and in the Cork Hospital, in only one year of the last forty has it slightly exceeded 6 per cent. ' In Barker and Cheyne's report of the Irish epidemic, 1817-19, it is stated that out of 100,737 patients in the hospitals of all Ireland, 4349 died, making the mortality 4·3 per cent., or only 1 in 23⅙. No doubt, as Dr. Murchison says, this small mortality is partly accounted for by the greater facilities afforded to mild cases for entering the hospitals in Ireland; but whether this be the case or not, it plainly shows that there is a form of fever constantly prevailing in Ireland, which is much milder, and in which, consequently, the mortality is much less, as compared with the fevers that prevail in this country.

The next table exhibits the mortality of the different forms.

TABLE VI.

Mortality of the different Forms of Continued Fever admitted into the London Fever Hospital during ten successive years, 1848-57.

YEARS.	TYPHUS.			ENTERIC.			RELAPSING.			FEBRICULA.	
	Admissions.	Deaths.	Per Cent.	Admissions.	Deaths.	Per Cent.	Admissions.	Deaths.	Per Cent.	Admissions.	Deaths.
1848	526	106	20·15	152	41	26·97	13	1	7·69	16	0
1849	155	39	25·16	138	26	18·84	29	0	...	79	0
1850	130	24	17·69	137	24	17·57	32	2	12·5	62	0
1851	68	6	8·82	234	30	12·82	256	7	2·73	56	0
1852	204	24	11·76	140	25	17·85	88	1	1·13	129	0
1853	408	90	22·06	211	59	27·96	16	0	...	152	0
1854	337	67	19·88	228	44	19·3	5	0	...	144	0
1855	342	83	24·27	217	30	13·82	1	0	...	62	0
1856	1062	207	19·49	149	23	15·43	...	...	...	89	0
1857	274	69	25·18	214	27	12·61	1	0	...	72	0
Total......	3506	715	20·36	1820	329	18·07	441	11	2·49	861	0

According to this table, the mortality of typhus, including all cases, is 20·39 per cent., or nearly 1 in 5; but if the cases that died within twenty-four hours (49) be deducted, it is 19·3 per cent. only; or if those which died within forty-eight hours (115) be subtracted, it falls to 17·7 per cent., or about 1 in 5⅔. This appears to be the average death-rate in other places. For example, in the Edinburgh Infirmary in the years 1848 and 1849 it amounted to 22·3 per cent., and in Glasgow Infirmary, for eleven years, it was 18 per cent.

The mortality of enteric fever turns out to be rather under that of typhus. Out of the 1829 cases treated in the London Fever Hospital, 333 died, being in the ratio of 18·29 per cent., or about 1 in 5¼; and if the cases fatal within twenty-four hours be deducted, it falls to 17⅔ per cent.; or if the deaths within forty-eight hours be subtracted, it is less than 17 per cent., or nearly 1 in 6. In one year, however, (1853,) the death-rate was greater than that in any year from typhus—viz., 28 per cent., or about 1 in 3½; and in another year (1848) nearly 27 per cent., or about 1 in 3$\frac{7}{10}$. The smallest mortality in any year was about 13 per cent., so that in no year was it so low as it has been in some years from typhus; and as has been pointed out by Dr. Murchison, the year in which the mortality was least, was also that in which there was the greatest number of cases, whereas the mortality from typhus appeared to be lowest when it was least prevalent.

In the Glasgow Royal Infirmary, the mortality from this form (enteric fever) has exceeded even that in London, being 21·6 per cent., or nearly 1 in 4½.

The mortality of relapsing fever is very small. According to the table it is not more than 2½ per cent., or 1 in 40. In the Edinburgh epidemic of 1843-40, it was under 4 per cent., and in that of Glasgow, between the years 1843 and 1853, it was 5 per cent.

It is important to keep in view, that the comparatively small rate of mortality of relapsing fever has an important bearing on the question of the death-rate of fever as a whole. In proportion as it prevails at any time, will the mortality of fever, including all forms, be diminished; and this circumstance explains the small rate of mortality in the London Fever Hospital in 1851, as well as in the epidemics that prevailed in Edinburgh, Aberdeen, and Glasgow in 1843.

Dr. Murchison strongly, and with much truth, urges the importance, when comparing the mortality from fever at different times and places, in order to judge of the merits of different plans of treatment, or for other purposes, of taking into account the form of fever which has prevailed. Thus, while the total mortality from fever in Glasgow was much below that of the London Fever Hospital, that in each of the individual forms was greater, the difference resulting from the much larger proportion of relapsing cases which occurred in Glasgow.   The same remarks apply also to febricula, no fatal case of which form was observed; but the greater the proportion of cases, the less will be the rate of mortality for all the cases of fever taken together.

It is very probable that the small death-rate constantly observed in the Irish fevers may admit of the same explanation.

The influence of age on the mortality of fevers is shown by the following table, which gives at one view the death-rate at each quinquennial period of life.

TABLE VII.

Mortality of the different Forms of Fever for each Quinquennial Period of Life, deduced from 6628 Cases admitted into the London Fever Hospital in Ten successive Years, 1848-57.

AGES.	TYPHUS. Admissions	Deaths	Per Cent.	ENTERIC. Admissions	Deaths	Per Cent.	RELAPSING. Admissions	Deaths	Per Cent.	FEBRICULA. Admissions	Deaths.
Under 5 years	17	3	17·65	4	0	0	4	0	0	8	0
From 5 to 10	183	14	7·65	103	14	13·59	32	0	0	83	0
„ 10 to 15	363	18	4·95	250	32	12·8	63	1	1·58	133	0
„ 15 to 20	546	26	4·76	519	84	16·18	92	1	1·08	186	0
„ 20 to 25	495	48	9·69	404	81	20·05	76	0	0	163	0
„ 25 to 30	343	52	15·15	240	45	18·75	37	0	0	60	0
„ 30 to 35	323	55	17·02	100	30	30	37	3	8.1	67	0
„ 35 to 40	270	89	32·96	60	14	23·33	19	2	10·52	45	0
„ 40 to 45	292	87	29·79	46	8	17·39	40	1	2·5	44	0
„ 45 to 50	212	82	38·68	20	5	25	8	2	2·5	20	0
„ 50 to 55	150	77	51·33	8	2	25	15	0	0	12	0
„ 55 to 60	100	51	51	9	4	44·44	7	1	14·28	9	0
„ 60 to 65	88	49	55·68	7	4	57·14	5	0	0	6	0
„ 65 to 70	42	29	66·66	1	1	100	1	0	0	5	0
„ 70 to 75	24	17	70·83	0	0	0	1	0	0	3	0
„ 75 to 80	6	5	83·33	1	0	0	0	0	0	1	0
Above 80 years	2	2	100	0	0	0	0	0	0	0	0
Not ascertained	50	12	22	49	5	10·41	4	0	0	16	0
Total	3506	715	20·36	1820	329	18·07	441	11	2.49	861	0

From this table it appears that the mortality of all forms from five to ten years of age is 7·65 per cent.; between ten and fifteen, it is less than at any other period of life; after this, it increases rapidly, until of those above fifty years of age, the mortality reaches 48⅓ per cent., or nearly one-half.

The death-rate from typhus between the ages of five and ten, appears to be only 7·65 per cent.; between ten and twenty it is under 5 per cent. After the age of twenty, however, it increases rapidly, so that of those above thirty years of age, the mortality is 36 per cent.; above forty, it is 43·66 per cent.; above fifty, it is 55·82 per cent.; and above sixty, it reaches as high as 62·34 per cent.

The mortality from enteric fever between the ages of ten and fifteen (about 13 per cent.), shows a smaller death-rate in early life than in typhus. At the different periods of life, too, there is greater uniformity in the mortality than in typhus, though, as in typhus, the death-rate keeps pace with advancing years. For example, above thirty years of age, the ratio is 27·38 per cent.; above forty, it is 27·17; above fifty, it is 46·15; above sixty, it is 55·55 per cent.

It may be noticed as singular that in both typhus and enteric fever, the death-rate is lower between the ages of forty and forty-five than in the period of life immediately preceding, nor should the fact of the greater mortality of typhus and relapsing fever in advanced life be overlooked.

Lastly, from the mortality table deduced from the records of the Fever Hospital, it comes out that the mortality from fever of all forms increases as life advances, the mean age of the fatal cases far exceeding that of those who recovered; for while the mean age of those who died (including all forms) was thirty-six, the mean age of those who recovered, was only twenty-four.

As to the influence of sex and of season on the mortality of fever, neither the one nor the other seems to have any particular effect.

GENERAL OBSERVATIONS ON THE TREATMENT
OF FEVERS.

Prophylaxis—change-of-type theory considered—management of the sick chamber.

BEFORE entering on the consideration of the treatment applicable to the several forms of fever, let me offer a few observations connected with their prophylaxis or prevention.

If the modern view of the nature of fevers—the doctrine which teaches us to consider all forms as originating in certain changes in the blood—be accepted, it is a natural subject of inquiry if there be any means of counteracting or neutralizing the effects of the fever poison, and thus preventing the development of the disease ?

By some, bloodletting has been recommended, though were such a proceeding in the present day hinted at, few patients would be inclined to submit to it. In former years, however, I have seen this treatment pursued, and in one of my own attacks, when certainly all the usual premonitory symptoms had apparently set in, I am satisfied that the disease was arrested by a single prompt bleeding; and I certainly have witnessed the same good effects from this remedy in others. But this practice was put in force in days when bloodletting was supposed to be the great remedy in all acute diseases, including fevers. The case would, however, be special indeed for which, in the present day, bleeding would be resorted to as a prophylactic or preventive remedy.

The exhibition of emetics has been proposed, on the supposition, that the virus or poison of fever acting primarily on the gastric system, the early administration of an emetic, by removing the offending agent, might possibly intercept its absorption into the blood. According to others, by the act of vomiting, and the general commotion thus induced, the poison has been supposed to be got rid of, and its effects prevented. Some, again, have thought, that as there is an admitted sympathy between the stomach and the skin, the sickness and vomiting cause a determination to the surface, and in this way an exit for the disturbing material may be established. But, unfortunately for this theory, the absorption of the poison is more probably effected through the pulmonary system, and with such rapidity does it often appear to act on the blood and solids, that even the early exhibition of an emetic may be too late for the anticipated results.

Equally insufficient grounds have been adduced for the employment of diaphoretics, which appear to have been suggested by observing the occasional termination of acute diseases by sweating. No one, however, I believe, ever witnessed a single instance in which there was the smallest ground for believing that the fever poison was dislodged by such means.

P

The cold affusion, first introduced and warmly advocated by Dr. Currie, has been put to the test of experiment; but notwithstanding the bold affirmation of its author, that if resorted to at the very commencement of the fever, at all events before the third day—it very generally, if not invariably, put an end to the symptoms, in fact, extinguished the disease—the experience of those who have given this method a fair and impartial trial is by no means favourable to its adoption. When I acted as clinical assistant to the late Professor Home in the Edinburgh Infirmary, the cold affusion was fairly tested, and the result was, that not in a single instance was the fever cut short, or its duration apparently abridged by it. It should, moreover, be kept in view that it is applicable only to the more acute forms of fever in vigorous subjects; for in the more delicate, the shock may be too great, and death even result from the practice— an event which occurred many years ago in the family of a late eminent professor, and greatly tended to throw the practice into discredit, and ultimately led to its final abandonment. Dr. Williams, in his work on "Morbid Poisons," remarks, "If we turn from this empirical practice of cold affusion to the great and leading doctrine of fever, which attributes this disease to the action of a morbid poison, it will be plain, if that hypothesis be correct, that cold affusion could not interrupt the course, though it might modify the symptoms.

A poison circulating with the blood cannot be re-moved from the system by ablution of its surface. No person expects to stay the course of small-pox, of scabies, or of syphilis by a similar application. We might, therefore, have predicated, *à priori*, that cold affusion could not remove from the body the poison of typhus fever, and consequently had no power to stop the course of the disease, though it might modify the symptoms."

The cinchona bark (or the salts of quinine) has been supposed to have a specific power in arresting the progress of malarious fevers. In the continued fevers of temperate climates, of whatever form, quinine has also been exhibited as a prophylactic in large doses, repeated at short intervals until the pe-culiar effect of the remedy—cinchonism—has been induced, indicated by headache, vertigo, tinnitus aurium, and a sedative effect on the heart's action. This remedy, like others, after being much extolled, had fallen into disuse, until recently revived by Dr. Dundas, of Liverpool, both as a prophylactic and curative agent. The united testimony, however, of those physicians of whom I have made inquiry as to the results of trials made with this remedy, leads to the inference, that its administration in any form of continued fever, as a preventive or prophylactic, cannot be relied on, and accords with my own experience in the cases in which I have employed it with this view.

From these cursory remarks, therefore, it is apparent that fever, when once developed, can rarely be arrested; or, in other words, that the means of expelling the poison, or of depriving it of its noxious effects, have yet to be discovered. The duty of the practitioner, therefore, is to endeavour to guide the disease, and to prevent as much as possible injury to organs essential to life, bearing in mind that this requires a certain and probably definite time to accomplish, even under the most judicious treatment.

Before I endeavour to lay before you the modes of treatment adapted to the various types of fever, I must allude to a subject of much practical import—the supposed recent change in the treatment, not only of fevers, but of acute diseases in general. It is a prevalent idea, that of late years every form of acute disease has assumed a less sthenic character, or to adopt the modern phraseology, has undergone a "change of type." This doctrine is now agitating, in no small degree, the medical and even the general community, and, as usual in such controversies, with no lack of combatants on either side. The one party contends, that the changes that have recently taken place in the views of a considerable proportion of the profession as to the treatment of acute maladies, including fevers, are to be ascribed mainly to the circumstance that the whole range of pyrexial diseases has of late—for more than a quarter of a century—undergone a marked change of type, or (to adopt the term employed by Sydenham) in epidemic constitution, the vital powers being supposed, from unknown or unexplained causes, to be in a low or depressed state, and therefore unable to withstand the combined influence of the disease and the routine antiphlogistic measures, more especially bloodletting. Others assert, and in my opinion upon more philosophic grounds, that the recent therapeutic innovations are the result of improved views of pathology, as well as of a more just appreciation of the powers of remedies. They entirely repudiate the idea that there has been such an alteration in the epidemic constitution or type of diseases, as to warrant the almost total change of treatment that has been recently advocated.

If we examine closely this theory as applicable to the acute diseases of the last thirty years—and this can only be undertaken by those who have witnessed and studied their type during the period referred to—and weigh dispassionately the evidence adduced, more especially by Nature herself, I apprehend that the true explanation of the difference of treatment will be found to consist in the more cautious and restricted notions now entertained as to the necessity for the heroic remedies formerly so freely, and I may say indiscriminately, adopted. Even those who are in favour of the change-of-type doctrine cannot assert that the pathological phenomena of acute maladies have undergone a change; nor have the symptoms general and local of the entire class of pyrexial diseases exhibited the slightest alteration; and if the evidence as to the depressed or asthenic condition of the vital powers be scrutinized, I have a strong impression that the conclusions adopted are not warranted by facts. And how important is it to form a correct judgment of this doctrine, since it determines the line of treatment to be pursued, and may even involve the safety of valuable lives!

The subject has not escaped my attention, and I am satisfied, that though certain cyclical differences in acute diseases, of longer or shorter duration, may have been occasionally detected, the notion of change of type, though applicable in some degree to fevers, has been greatly exaggerated.

Let me state the grounds upon which I have come to this conclusion, which I admit is at variance with the ideas of many physicians whose knowledge and judgment entitle them to great consideration.

We find that Sydenham, who is considered to be the author of this change-of-type theory, cautioned the medical men of his day against too hastily determining the treatment of a new epidemic—until, in short, the practitioner and the disease were better acquainted—on the reasonable ground, that

epidemics assumed at one time a more acute or
phlogistic, at another a less acute or asthenic character. But we do not find that he had observed a
change in one direction only, and for so lengthened
a period as considerably more than a quarter of a
century. If the records of epidemics of other
febrile diseases—the exanthematous fevers, for example—be traced, it will be evident that, during
the same period (in small-pox, measles, and scarlet
fever), every variety or modification of type has
been observed, the type being sometimes acute,
sometimes more or less asthenic, and requiring,
consequently, variation in treatment.

But I am strongly persuaded that, in regard to
fevers, the true explanation will be found in the
fact that, until very recently, little or no attention
has been paid to the ever-varying differences in
form which they assume—at one time typhus, at
another enteric (or typhoid), or it may be relapsing fever—constituting the features of the
prevailing epidemic. So that the question of the
identity or non-identity of the several forms of
continued fevers becomes of the greatest importance in relation to the change-of-type theory.

The great argument adduced by those who support the doctrine (change of type) is, the favourable results in the Edinburgh epidemic of 1817-20
(which I had the opportunity of witnessing) of
large indiscriminate bleedings in diminishing the

mortality. We are told, somewhat exultingly, that under the unnecessarily profuse phlebotomy, the mortality did not exceed 1 in 22 at any period of the disease, and was reduced so low as 1 in 30 as the epidemic spread. This argument, however, loses much of its intended effect, when it is considered that by much the larger number of cases consisted of relapsing fever—a form the mortality of which has already been shown to be exceedingly small under opposite modes of treatment, and in which the death-rate has been even less when no blood was abstracted at all. For example, in that of 1843, the history of which has been given by Dr. Cormack, the death-rate was 1 in 16; of the cases recorded by Dr. Wardell (1843-4), it was 1 in 20; and of 203 cases treated in the Edinburgh Infirmary in 1848-9, there were only 8 deaths; and if we extend our inquiries to other places, we find that of 7804 cases of relapsing fever admitted into the Glasgow Infirmary between the years 1843 and 1853, the mortality was 405, or about 5 per cent.; and in the London Fever Hospital, of 441 cases admitted during ten years (1848 to 1857), 11 died, being in the ratio of about 1 in 40.

This variation in the mortality could not be ascribed to the measures employed; for Dr. Cormack states that, having been urged by medical friends to test the effects of bloodletting, he instituted trials of this remedy, but candidly ad-

mitted that, though the symptoms were sometimes
evidently relieved, the beneficial changes were
often not effects, but sequences of the bleeding, as
was satisfactorily proved by the very same changes
frequently occurring, as suddenly and unequivocally, in patients in the same wards, and affected
in the same way, who were subjected to no treatment whatever. And in regard to the measures instituted at the London Fever Hospital,
when the mortality of relapsing fever did not
exceed one in forty, with scarcely an exception,
blood was not abstracted at any period of the
disease.

It is evident, therefore, that the change-of-type
theory cannot rest on comparison of the treatment
by indiscriminate phlebotomy formerly practised,
when all acute diseases, including fevers, were
supposed to be under the dominion of the lancet.
But though the grounds on which this question
has been argued are, in my opinion, erroneous, one
good result has followed—the death-blow which the
practice of indiscriminate bloodletting formerly
adopted in acute maladies has received; for too
often, little or no regard was paid to individual
peculiarities, or even to the stage of the disease for
which the bleeding was employed. The inquiry
was simply as to the existence of fever or of inflammation; and, the question once settled, the
lancet was unsheathed, and much blood unneces-

sarily shed, from the effects of which the patient did not recover perhaps for months. On the other hand, there is great hazard of many important diseases being allowed to gain the ascendancy, in consequence of the indecision that has resulted from the complete alteration of therapeutic principles which the discussion of this question has brought about. My own experience tells me, that I have witnessed the same good effects from the moderate abstraction of blood in some forms of acute disease—occasionally even in the early stage of enteric fever—as in by-gone days. I have not observed that loose condition of the crassamentum of the blood, nor the absence of the buffy coat, so characteristic of low vitality; but I have been careful in the selection of cases, as well as of the period of the disease, before resorting to bloodletting, and I am satisfied, that with this precaution, there will in general be little regret that this remedy, powerful for good or for evil, has been had recourse to.

It is, however, consoling to observe, that in the present day there is a more just appreciation of the powers of curative agents, as well as of the principles on which they should be applied in the treatment, not only of acute, but of chronic maladies. We are now, or ought to be satisfied, that the most scientific, as well as the most successful course in many diseases, is, after a certain

period, not to interfere too much, if at all, with
the operations of Nature in her efforts to repair
the injury parts or organs have sustained by
disease.

There are one or two other points to which I
deem it necessary to allude.

And first, let me remark that though the pathology of this class of diseases (fevers) has been
minutely investigated, their therapeutics, or the
principles on which the treatment should be conducted, have not received corresponding attention.
This is to be ascribed, in some measure, to the
circumstance, that many (especially of the Culle-
nian school), still hesitate to accept, if they do
not altogether deny, the distinctions between the
different forms of continued fever, and consequently apply indiscriminately the same therapeutic indications to all. But those who have had
much experience in the treatment of fevers are
satisfied that the curative principles of each form
differ in essential particulars.

Again, it must be obvious, that before deciding
on the measures to be adopted in a given case,
— of whatever type or form, — several circumstances should be kept in view: for example, its
mildness or gravity, its stage or duration; the
existence of secondary lesions; the prevailing or
epidemic constitution—that is, the varying but unknown causes which influence acute diseases in

general; and lastly, the individuality of the patient, if I may use the expression, including the age, habits, occupation, condition of life, previous state of health, and such collateral information as can be obtained from authentic sources.

Next let me say a few words on the manage-ment of the sick chamber—a point of no incon-siderable importance, though too often overlooked, and closely connected with the comfort of the invalid, and often with the issue of the case.

The first consideration should be the selection of a properly qualified nurse. The value of the ser-vices of one accustomed to fever cases can scarcely be over-estimated, as, besides capacity for receiving general directions from the medical attendant, there are numerous details that must be left to her dis-cretion, and which can only be satisfactorily under-taken by a person of judgment and experience, and who has been well trained for such duties. In severe cases, a nurse for day and another for night duty is indispensable.

We shall presume that the apartment is as large as is necessary for the requirements of the patient. The ventilation as well as the temperature of the room should be attended to. The former may be effected, not by keeping the window constantly open, often regardless of season or weather, but by allowing a current of fresh air to pass through the room occasionally, and then closing it. The temperature should be under 60°, though sometimes, in the summer and autumn months, it may be difficult to obtain this limit. Those who have had much experience of the management of fever among the poorer classes are aware of the rapidly favourable change in the symptoms, after removal from ill ventilated and crowded dwellings to the spaous wards of an hospital. The whole complexion of the disease, indeed, is often improved in a few hours.

The bed and body linen of the patient should be frequently changed, and the soiled linen removed at once from the sick room and immersed in cold water. The skin should be sponged with cold, or in winter with tepid water, to which a small portion of vinegar is added, once a day, or more frequently if there be irregular accessions of feverish heat. Sometimes it is sufficient to sponge the palms of the hands, an operation always grateful to the patient.

The bed on which the patient is placed should be of moderate size, and furnished with a soft hair mattress; and if the room be of sufficient dimensions to contain two small beds, the comfort of the invalid will be promoted by occupying one during the day, and the other in the night. This plan is indispensable in the more severe cases, and more especially so, when the evacuations are passed unconsciously. The room should be kept perfectly quiet, the light subdued, and only the attending nurse, and an occasional judicious visitor, when sanctioned by the medical attendant, allowed to have access. The visits of such persons should be short.

TREATMENT OF ENTERIC FEVER.

Of the mild form—of the more severe—question of bloodletting con-
sidered—treatment of the intestinal affection—diarrhœa—in-
testinal hæmorrhage—tympanitis—dietetic management—treat-
ment of the advanced period—wine and stimulants—treatment
of secondary affections—of the brain—of laryngeal angina—of
the pulmonary complications—of peritonitis and intestinal per-
foration.

PRESUMING that provision has been made for the
due administration of our remedies, and that the
form and stage of the fever, with the individual
circumstances of the patient, have been ascertained,
we shall suppose that we have to deal with enteric
fever of a mild character and in the early stage.

I need scarcely observe that the patient should
be confined to bed, since this part of the management is generally consonant with the feelings,
though sometimes, under the impression that the
disease is of a trifling nature, an effort is made to
keep about until the increasing prostration compels
the patient to remain at rest.

Emetics are occasionally employed, but only at
the very beginning of the disease, and when there
is evidence of the presence of acrid matters in the
stomach. Under such circumstances, the exhibition of an emetic, though it may not cut short the
fever, is often followed by much relief. But the
opportunities of prescribing emetics are comparatively rare, as medical aid is seldom sought at the

very beginning of the disease, to which period their administration should be restricted. As to the supposed benefit from the shock given to the system, or the determination to the skin or other organs induced by the act of vomiting, I have already confessed my doubts, and I should hesitate to adopt or sanction their use on such grounds.

In regard to aperients, I may observe that they, too, should not be prescribed indiscriminately in enteric fever. The bowels generally act spontaneously, and if they do not, the mildest aperients should be selected, merely with the view of relieving the system of any accidental accumulation. But let me caution the inexperienced in the management of this form of fever against the calomel and black draught system, which may do incalculable injury, by irritating the bowels and increasing the local intestinal disease.

No dependence should be placed on salines, as they are called, but, if something must be done to amuse the patient, a julep of the citrate of potash or ammonia is very grateful, and may tend to allay thirst and promote determination to the skin. The German Seltzer water is also a very grateful, harmless saline.

It may be inquired, what is the condition of Peyer's glands in these mild cases of enteric fever? They are probably simply enlarged from the deposit of typhous matter, but the subsequent changes are arrested, and they no doubt return to their

natural state as the general febrile movement subsides and convalescence takes place.

In such cases, therefore, medical interference
should be restricted to the management of the sick
chamber, enjoining mental and bodily repose, re-
gulating, if necessary, the alvine excretions, and
restricting the diet to simple farinaceous matters
and any of the weak animal broths. If it be deemed
necessary to prescribe salines, an occasional dose of
alkaline citrate may be interposed.

In by far the larger proportion of cases, however, enteric fever assumes a much more severe
character. There is more violent febrile disturbance from the beginning, indicated by more
frequent and severe rigors, and pungent and enduring heat of skin; while the condition of the
vascular, nervous, and gastric systems shows that
the poison has acted with considerable intensity.
The disease, in short, is of a more serious character, and consequently requires more decided
measures, modified, of course, according to the
circumstances each individual case presents. If
there be much general excitement, more especially
throbbing headache, flushing, restlessness, and perhaps delirium, the vascular fulness must be reduced. This may be accomplished by small doses
of antimony, repeated every three or four hours,
shaving the head, and applying a spirit lotion to the
scalp. Should this plan not succeed in combating

the cerebral excitement, or should symptoms in other organs arise to indicate that the struggle is to be severe, the question of bloodletting is forced on our consideration.

It would be unprofitable to discuss the various grounds on which bloodletting has been employed in the treatment of fevers. I would rather consider the subject as one to be determined by experience and observation; and it is to be regretted, that in our own country the effect of bloodletting as regards enteric fever has scarcely been put to the test, this remedy having been employed indiscriminately, without reference to the type or form an epidemic may have assumed. On the European continent and in the United States, where enteric fever is endemic, we have the recorded opinions of practical physicians; and, as might be anticipated, there is much difference in their views as to its utility, the remedy being extolled by some, condemned by others, and in the present day, if not entirely abandoned, adopted under very special circumstances only.

If we refer to Louis, we find that he abstracted blood in nearly all his cases, but restricted the practice to the early stage of the disease, and repeated the bloodletting according to the severity of the symptoms and the powers of the patient. In his opinion it was useful, within those limits, in shortening the duration and mitigating the violence of the fever; but in mild cases, even

in the early stage, as well as in the more severe,
after the second week, he was satisfied that the
abstraction of blood was of doubtful efficacy, if not
positively hurtful.

Andral does not mention the practice with
favour. He prescribed it in seventy-four cases, of
whom thirty-five—nearly one-half—died; and of
this number, in sixteen only, was there evident
amendment in the symptoms; while in sixty-four
the disease appeared to be even somewhat aggravated by the loss of blood. But it would have been
more satisfactory had the description of the symptoms for which the bleedings were employed, the
age and condition of the patients, and especially
the period of the fever, been detailed.

Chomel, who was a thoroughly practical physician, thought on the other hand, even in mild cases,
a single moderate bleeding at the commencement
beneficial in relieving the more urgent symptoms,
shortening the duration of the fever, and preventing local affections. In the more severe, in which
there was unusual general excitement, and especially if there were serious affection of some organ,
he advised the venesection to be repeated; but he
enjoined caution even under such circumstances,
and moreover warned the practitioner that a smaller
amount of blood should be abstracted than in
primary inflammation.

The great advocate of bloodletting in this disease,
however, is Bouillaud. This distinguished Pari-

sian professor, in his "Philosophie Médicale," an-
nounced that he had discovered an infallible mode
of treating all acute diseases, including fevers, by
copious bloodletting. He prescribed five or six
bleedings in succession, to the extent of twelve or
sixteen ounces at a time, as much being often
abstracted by cupping or leeching in the intervals;
nor were these heroic measures limited to the early
stage, but adopted even in the second week of the
fever. This bold treatment gave rise soon afterwards to much discussion in the Parisian medical
circles, which was carried on with no little warmth,
and led to a rigid examination of the grounds on
which its boasted success was founded. The investigation was instituted by Louis, whose treatment had been placed in unfavourable contrast,
and he had little difficulty in exposing the fallacy
of the statistics of Bouillaud, as well as the diagnostic errors into which he had fallen.

In the United States, where, in consequence of
enteric fever being epidemic, ample scope is afforded
for observing its pathology and treatment, it would
appear that the practice of indiscriminate bleeding
does not find favour, its employment being restricted to special circumstances; as, for example,
when the febrile symptoms run high in young and
vigorous subjects, but more especially, if there be
severe headache, flushing, throbbing in the cerebral
vessels, or acute delirium. For such symptoms
the physicians of the United States do not hesitate

to prescribe either general or local bloodletting, according to the emergency.

I have not brought these opinions before you with the object of recommending bloodletting in the treatment of enteric fever as *a general rule.* On comparing carefully the results of cases treated by others with my own experience, I am satisfied that in mild cases, loss of blood is uncalled for, and, if practised, tends only to lower the powers, and retard convalescence; that in cases of severity, and especially when the symptoms pursue a downward course, indicated by marked prostration, compressible pulse, low delirium, or exhausting diarrhœa, every ounce of blood abstracted diminishes the chance of recovery. When the fever assumes this grave aspect, there are influences at work that produce a depressed state of the vital powers, rendering evacuations of any kind, and especially bloodletting, hazardous in the extreme.

But intermediate cases are of not unfrequent occurrence, in which, in addition to the intestinal affection, there is unusual general excitement, or it may be, that some organ important to life is implicated. Under such circumstances, I have observed marked relief after the loss of a few ounces of blood, (rarely exceeding ten), if abstracted in the early stage of the fever; and in the comparatively few instances in which, of late years, I have ventured on this remedy, it has been apparent that, besides the improved feeling admitted by the patient, the duration of the fever was shortened. In doubtful cases, in which the propriety of bloodletting may be questionable, antimonial preparations are of great service in subduing the excitement without the hazard of lowering unduly the powers of the patient.

We have seen that the poison which induces enteric fever has a specific action on the intestinal glands (Peyer's patches), which undergo the changes formerly explained. This intestinal affection is one of the earliest phenomena, and accounts for the diarrhœa which is so commonly present from the commencement of this fever.

I must, however, protest against the practice occasionally adopted of treating the intestinal affection by local abstractions of blood, as if it were a local phlegmasia—a practice which, without improving the intestinal disease, tends only to impair the patient's strength. The typhous deposit in the intestinal glands, not strictly secondary, bears the same relation to the fever as the variolous eruption does to small pox, or the morbillous rash to measles. Should, however, the follicular disease produce great irritation, or, perhaps, inflammation of the mucous membrane, or should the morbid action spread deeper and implicate the peritoneum, no time should be lost in applying leeches over the region of the cæcum, followed by warm poulticing, and a combination of mercury with chalk and Dover's powder, at short intervals.

In the management of the diarrhœa, let me offer you a caution against the exhibition of aperients, which some French physicians—even Louis himself—have recommended.

In the very commencement, and particularly if there be uncertainty as to the state of the bowels, it may be well to exhibit a mild laxative, such as castor oil, magnesia, phosphate of soda, or rhubarb; and should the selected aperient act unduly, an aromatic draught, with a few drops of laudanum, may be given to check the irritation.

If the diarrhœa be moderate, it should not be interfered with, but treated simply by bland fluids, such as thin barley-water (made by infusion), rice-water, arrowroot; weak animal broth, in moderate quantity, being at the same time allowed.

If it be necessary, for effect, to prescribe, the almond mixture, or a mixture containing sperma-ceti, with or without a few drops of laudanum, or compound tincture of camphor, may be given.

When the evacuations are frequent (more than three or four in twenty-four hours), the powers of the patient become enfeebled, not only by the drain on the system, but the very exertion in getting out of bed tells on the patient's strength. It becomes, therefore, necessary to check this symptom. Some-times a few drops (five to ten) of laudanum, or fifteen to twenty of the compound tincture of camphor, in any agreeable vehicle, answers very well, and may be repeated according to circumstances. The local exhibition of opium by an enema, containing ten to fifteen drops of laudanum in three or four ounces of barley-water, or thin starch-gruel, and repeated according to circumstances, is often preferable.

But if, notwithstanding the use of opiates in the way mentioned, the diarrhœa persist, we must have recourse to other remedies of the stringent kind, either alone, or combined with opium. We may select from the mineral or vegetable kingdom, the former affording, however, the best and most suitable.

The mineral acids—the diluted sulphuric, nitric, or phosphoric—combined with laudanum in the proportion of fifteen minims of the one (the acid) to three, four, or five of the other (the laudanum), are well suited, and may be taken at longer or shorter intervals, according to circumstances.

The diacetate or sugar of lead has considerable power over the intestinal affection. I am in the habit of giving it, even in the early stage, under the impression, from its power in checking hæmorrhage from mucous surfaces, that it is capable, not only of controlling the diarrhœa, but of keeping in check the ulcerative process in Peyer's patches. It may be given alone in three-grain doses, either in pill or solution, or in combination with small doses of the solution of acetate of morphia, every six or eight hours, or after every evacuation. The solution of acetate of morphia (made by dissolving six grains of the alkaloid in two drachms of spirit of wine and six of distilled water), should be used in preference to laudanum, with which sugar of lead does not appear to harmonize.

The sulphate of copper in combination with opium is a valuable remedy. Indeed, I do not know one better suited, especially in protracted cases. A pill, containing an eighth of a grain of this salt, and the same quantity of solid opium, administered every four, six, or eight hours, will seldom fail to check the diarrhœa, while there is reason to suppose that, like sugar of lead, it exerts a beneficial influence on the intestinal affection.

From the nitrate of silver I have also seen excellent effects. It may be given in quarter-grain doses, with or without opium, every six hours, or after every liquid evacuation. The possibility of its darkening the skin is, however, an objection to its employment urged by some practitioners; but though I have prescribed it extensively in this intestinal affection, and continued its use for a considerable time, I have never witnessed any approach to discoloration, nor do I think, in the small doses recommended, there is much risk of it.

Alum is another remedy of the astringent class to be relied on, the best mode of administering it being in the form of alum whey, made by throwing one drachm of alum in powder into a pint of boiling milk, the whey being separated from the coagulum by straining. Of this two tablespoonfuls may be given alone, or in combination with compound tincture of camphor, at intervals of three or four hours, or after every liquid evacuation.

Of remedies from the vegetable kingdom, I may mention krameria, logwood, and catechu, which, like the mineral astringents, are most efficacious when combined with opium.

I may, however, remark that, on the whole, I have seen most permanent benefit from sugar of lead, sulphate of copper, and nitrate of silver, in combination with opium, either of which, when judiciously managed, will do all that an individual remedy can effect in controlling the diarrhœa.

Besides internal remedies, various local applications often afford material assistance in the treatment of the intestinal affection. Much benefit is sometimes derived from warm applications—either a large, thin linseed poultice, or, what is a more agreeable fomentation, two folds of lint, sufficiently large to cover the belly, dipped in warm water and applied quite hot, the moistened lint being protected by an oil-silk covering to prevent evaporation. This may be renewed from time to time, according to the feelings of the patient, and is always grateful and soothing. If the tympanitic distension be considerable,—and there is more or less in every case,—oil of turpentine may be added to the warm poultice or moistened lint, and kept on as long as it can be borne, the tolerance of this stinging appli- cation varying with the sensibility of the skin. I am in the habit of keeping the right iliac region tender by the turpentine, when its more extended application over the abdomen is either unnecessary, or objected to by the patient. If it produce much pain or irritation of the skin, it should be used occasionally only—indeed, few patients can bear it long at a time. If the odour of the turpentine or the burning pain it frequently induces, be an ob- jection, the compound camphor liniment may be substituted.

Hæmorrhage from the bowels is a pretty certain indication of intestinal ulceration, the source of the blood being principally from the ulcerated sur- face. I have often been surprised at the large quantity discharged without the patient being ma- terially lowered. This happens, however, only in robust subjects, and in such circumstances we must not hastily interfere with this symptom. But those who have less constitutional vigour, are generally much enfeebled by even a moderate dis- charge of blood from the bowels, especially if it take place in the latter stage of the fever, so that in such cases it is necessary to adopt measures to arrest the hæmorrhage.

The most perfect quietude should be enjoined, cold or iced drinks given, the abdomen covered with cloths dipped in cold vinegar and water, and sugar of lead, in five-grain doses, with a few drops of the solution of acetate of morphia, or a combination of gallic acid with opium, given at short intervals. When the hæmorrhage is moderate, the starch-and-laudanum enema will often alone control it.

I have prescribed occasionally, with benefit, a nostrum called Ruspini's tincture. It consists principally, I believe, of gallic acid in some colouring fluid. The oil of turpentine, in doses of from twenty to thirty drops in emulsion, is also a good remedy, taking care to withdraw it, if it produce irritation of the urinary organs.

Should the powers be much reduced by the loss of blood, wine, brandy, and other cordials, are, in addition to the remedies pointed out, to be administered; while the nervous system is calmed by opium or its alkaloids.

The mucous membrane of the stomach and bowels has the property of secreting air in abundance during health, and when it takes on a congestive or inflammatory action this secretion is increased, and hence we find tympanitis an almost invariable accompaniment of enteric fever, more particularly in the middle and latter stages. The tympanitic distension of the belly is often distressing. It is sometimes so great as seriously to embarrass the breathing by impeding the descent of the diaphragm; and though the secreted air accumulates in great abundance in the colon, it is seldom passed per anum.

This symptom, if it occur early, and the powers of the patient are not much impaired, is best treated by the application of leeches to the abdomen, followed by emollient applications and the internal exhibition of the diacetate of lead. But should local bloodletting be inadmissible, turpentine stupes may be substituted, while the lead is given internally. It is often necessary, however, from the amount of distension, to adopt some mechanical expedients to get rid of the accumulated air, such as the introduction into the rectum of a flexible tube, or catheter of the largest size, or the common œsophagus tube of the stomach-pump, properly secured, so as to prevent it slipping into the bowel. But before having recourse to mechanical means, I am in the habit of prescribing a fetid enema, consisting of half an ounce or more of the tincture of assafetida in a pint of common gruel, or half an ounce of the confection of rue in camphor mixture. I have seen either of these remedies, carefully administered, followed by the expulsion of an enormous quantity of confined air, to the patient's great relief. If they fail, I have applied in some cases with happy effects the stomach-pump, per rectum, and withdrawn by the syringe the accumulated air, which may be passed through the lower tube of the stomach-pump into a basin containing water. But this should only be had recourse to when the introduction of the catheter or œsophagus tube fails.

In protracted cases, accompanied with gradual

emaciation and irregular hectic paroxysms, there is little question that the intestinal ulceration is in a state of slow activity; under such circumstances, I know no remedy equal to the sulphate of copper with opium, or the nitrate of silver (as already recommended), perseveringly administered, supporting at the same time the patient's strength by light broths, eggs, farinaceous preparations, and perhaps the more delicate kinds of white fish.

If there be not speedy amendment, the case is apt to be given up as hopeless; but the duty of the practitioner is to persevere, even under discouraging circumstances. Many a time have I witnessed patients, emaciated almost to a skeleton, gradually recover, and reward many weeks' anxious care, and even when it was thought scarcely possible that the intestinal ulceration would ultimately cicatrize.

Let me draw your attention to the bladder, which is liable to become distended in the advanced stages of this fever. There may be complete retention of urine, or it may partially escape unconsciously, and thus the nurse, or even the medical attendant, fancy all is right so far as micturition is concerned. There can be no difficulty, however, in determining the point, if the region of the bladder be carefully examined, a single percussion stroke being sufficient, to the practised finger, to indicate the presence of accumulated urine, unless there be such an amount of tympanitic distension as to render it impossible to come to a decided conclusion. If

there be doubt, introduction of the catheter should
not be delayed. In all cases, therefore, where
there is great prostration, more especially if there
be secondary brain affection, the state of the bladder should be daily inquired into, and not left to
the report of the nurse, however experienced and
attentive.

A point of much importance in the treatment of
enteric fever is the due regulation of nourishment;
indeed it requires equal discrimination as the administration of remedies strictly so called. The
practice formerly adopted of almost entirely pro-
hibiting nutriment in the management of fever
originated in the erroneous views entertained by
the solidists of the great similarity between, if not
the perfect identity of, inflammation and fever; and
with such ideas of its nature, we cannot wonder at
the dietetic restrictions inculcated. But whatever
difference may still exist as to the indiscriminate
exhibition of stimulants, few dispute the advantage
of giving suitable support even in the early stage
of enteric fever.

For the first few days, the most simple articles
only should be allowed, and thirst allayed by rice or
groat gruel, the natural Seltzer water, thin barley-
water, rennet whey, or weak black tea. The appetite is usually so much impaired that there is little
difficulty in imposing restriction, but as the fever
progresses, it is better to allow more nourishment,

in the shape of thin panada, with a small quantity of beef-tea or any of the animal broths. I generally allow from half a pint to a pint of moderately good beef-tea in the twenty-four hours, almost from the commencement of the fever, unless the symptoms indicate a more than ordinarily acute disease, and consequently a more restricted regimen. The beef-tea should be given in small portions at a time, and if it produce uncomfortable feelings or feverishness it should be put aside for a few days, and gruel and panada substituted.

In enteric fever, fruits and subacid drinks should be either almost entirely forbidden, or taken very sparingly, from the risk of increasing the irritability of the bowels; and while speaking of drinks, I may mention that the large amount of fluids taken with the object of checking thirst often produces uneasiness and feeling of oppression from the mere mechanical distension of the stomach, and very often without allaying it. It is better that small quantities of fluid, or a little piece of ice, be taken now and then, than copious draughts gulped down without reference to the uneasy sensations that are likely to follow; and for the same reason, soda-water, and such like drinks containing fixed air, should be avoided. The natural Seltzer water is unobjectionable.

Towards the middle or end of the second week, sometimes earlier sometimes later, symptoms indicative of exhaustion appear. The pulse becomes soft and oppressive; the skin cool, and often covered with

clammy moisture; the patient feels weaker, less able
to exert himself, or perhaps even to turn in bed,
the tongue sometimes assuming a brown appearance.
The necessity of a change of measures now becomes
apparent. When the fever has been mild, it is
generally sufficient to allow a more sustaining diet,
stronger soups, or a small quantity of the more
delicate kinds of white fish, or of panada of
chicken or game, bearing in mind the irritable
state of the bowels. At the same time a light
tonic may be prescribed, those from the vegetable
kingdom being by many preferred. The barks of
cinchona and cascarilla, the roots of gentian, chy-
retta, senega, calumba, serpentary, either in the
form of infusion or tincture, may be selected at
discretion, though the salts of quinine have superseded them in a great measure. Many even in
the present day adhere to the old decoction of
bark, under the belief that this preparation contains
more of the real principles of the drug than the
modern alkaloid. Battley's liquor cinchonæ, in
fifteen to twenty minim doses, is an agreeable and
elegant preparation. The mineral acids, too, are
also well suited; for example, a combination of
the diluted nitric and muriatic acids in infusion
or tincture of orange-peel. I have lately, on the
suggestion of Professor Huss, given the diluted
phosphoric acid, but I am not sensible of its having
any advantage over the others.

In regard to the administration of wine and

alcoholic fluids in the treatment of this stage of
enteric fever, I would simply observe, that they are
too frequently prescribed on the mere dogma of
routine; and even when the symptoms warrant
their employment, a much smaller quantity is required than in typhus fever. Many patients pass
through enteric fever without a single dose of
wine, even where the disease is protracted, showing
that its indiscriminate use is by no means warranted; and it should be kept in mind, that in its
progress there are, besides exhaustion, other processes at work in the system—local affections—
which possibly may be ultimately the destroying
cause, and these may be aggravated by the injudicious use of stimulants. On the other hand, it
is often imperative to support the failing powers,
disregarding the injury particular organs may have
sustained—in short, to obviate the tendency to
death,—and to hope that, as the constitutional
powers are invigorated, the secondary lesions will
partake of the general improvement.

But special circumstances sometimes arise, requiring the administration of stimulants, without
regard to the period of the fever or secondary
affections. The powers may suddenly give way,
rendering immediate and powerful stimulation necessary; and none is more suitable, under such
emergency, than brandy in half-ounce doses, repeated at such intervals as may be necessary. Again,
sometimes a few ounces of wine, or a proportional

amount of brandy, are sufficient to resuscitate the failing powers, and when this is accomplished, the remedy should either be withdrawn, or given occasionally only.

To sum up the question of the administration of wine in enteric fever, I would observe—1st, that it should not be prescribed indiscriminately, but as an occasional remedy in special cases; and 2nd, when there is doubt as to the propriety of giving or withholding wine, if the case be one of well-marked enteric fever, especially in the early stage, it will be safer to withhold it, at all events for a time; and 3rd, in every case, but particularly when wine is not clearly indicated, the effect of the first few doses should be closely watched, and if the general complexion of the symptoms be improved, it may be continued with care; if, however, it excite the patient, if restlessness and delirium be induced, if the tongue become more and more dry, the pulse more rapid and wiry, you may infer that wine is not suited to the case—at all events, at its present stage—and it should therefore be withdrawn.

Let me also press the importance of giving the wine at stated intervals, and only when the excitement is moderate. It is especially necessary to give it during the night, when there is often great exhaustion. A dose of wine judiciously given at this diurnal period is frequently followed by calm, refreshing sleep; and hence the incalculable advantage of an interested, experienced nurse, on whom so much responsibility — indeed, the life of the patient—often rests.

We have seen that in the progress of enteric fever the poison acts with more or less intensity on individual organs, inducing the so-called secondary affections or complications, and requiring special local treatment. It should, however, be kept in view, that these lesions, which are due rather to congestion than to inflammation, neither require nor bear the measures that are usually employed in primary or idiopathic inflammation.

The nervous system, on the whole, perhaps, suffers more severely than any other organ, and this is what might be anticipated from its seldom escaping in the general febrile commotion, except in the mildest cases. When, therefore, the cerebral affection becomes unduly prominent, indicated by constant pain in the head, especially if accompanied by flushing, delirium, intolerance of light, and sleeplessness, the loss of a few ounces of blood by cupping the temples, or the application of leeches behind the ears, should not be omitted, while the hair should be removed from the head so as to allow a spirit lotion, or iced water, to be applied. The topical bleeding, however, is advisable only at the onset, or first appearance of the brain symptoms, and when the powers of the patient are able to sustain the loss of blood. In this acute cerebral affection, some have recourse to blistering the scalp. I enter my protest against the practice, at all events

until the brain excitement has passed away, and even
then the effect of vesication may be questionable.

A more suitable and effectual remedy is the affusion of cold water on the scalp. It is generally followed by marked relief, and is more especially to be
employed when the powers of the patient render
topical bleeding doubtful or inexpedient. The
mode of application is as simple as it is efficacious.
The patient is raised in bed, the head supported
over an empty basin, and cold water poured over
the scalp, the stream being gradually raised till it
fall from a height of eight to twelve inches. A
considerable shock is thus produced, but it is
usually followed by such amelioration of the symptoms that its repetition is often called for by the
patient. It is exceedingly useful when the cerebral
excitement threatens to recur, and it should always
be had recourse to in severe cases as an auxiliary to
the other measures. When the brain affection assumes a form closely resembling delirium tremens
—the patient being sleepless, the pulse rapid, soft,
and compressible, the skin cool, the face pale, the
delirium of the low, muttering character, with
tremors and starting of the tendons—loss of blood
would speedily annihilate the chance of recovery.
The treatment for this condition of the brain consists of fomentations of warm vinegar and water to
the head; the application of blisters to the temples
and forehead (avoiding the nape), and sinapisms
to the extremities, supporting the strength at the

same time by wine and nourishment, while the
nervous system is tranquillized by a combination of
antimony and opium in small doses, repeated at
short intervals, according to the effect produced.
For the introduction of this combination in the
condition of the brain referred to, the profession is
indebted to the late Dr. Graves, who employed it
with the most happy results, and I am enabled to
confirm his statements from my own experience.
The mode of prescribing it is, to add to eight
ounces of a julep of the citrate of ammonia, four
grains of tartar emetic, and one drachm of the
solution of the acetate of morphia, of which half an
ounce is to be administered every two hours, until
sleep is procured. The scalp should at the same
time be enveloped in a spirit lotion, and care taken
that during sleep nourishment is administered at
regular intervals, and not postponed until the
patient awakes spontaneously. The repetition of
the remedy in the way recommended must, of
course, be left to the judgment of the medical
attendant and the exigency of the case.

Some practitioners prefer in these cases the local
administration of opium by enema; and no doubt
this mode is well adapted, as it tends not only to
calm the nervous system, but at the same time to
check the diarrhœa so commonly present.

Of the throat affections, the most formidable is
the laryngeal angina formerly alluded to. This
dangerous disease, which those who adopt the

views of Rokitansky ascribe to the typhous deposit in the delicate structure of the glottis, is, happily, not frequently met with, for it is seldom arrested by even the most prompt treatment. The first approach of hoarseness, with painful deglutition and tenderness on pressing the region of the larynx externally, should excite suspicion, and no time ought to be lost in employing such remedies as the local symptoms require, having always in view the powers of the patient and the probable endurance of the treatment. Gargles only produce fatigue and disappointment. Cupping on the nape of the neck, rapidly and dexterously performed, generally affords relief; and this should be followed by the application of the acetous solution of cantharidin under each angle of the jaw, avoiding the integuments covering the larynx; and, if the patient has strength for the effort, the vapour arising from hot water containing extract of conium should be conveyed to the throat by means of Read's valve inhaler. A large emollient poultice should be applied from ear to ear, and renewed every three hours or oftener. As to internal remedies, the exhibition of mercury, so manifestly efficacious in idiopathic angina of the larynx, is less trustworthy in the secondary form now under consideration, in consequence of the hazard of its irritating the bowels, and consequently renewing or increasing the local intestinal affection. If it be employed at all, the safer mode is the application of mercurial ointment to a portion of integument—the nape of the neck, for example—from which the cuticle has been removed by a blister. But a more safe and efficacious remedy will be found in antimonial preparations, especially the tartar emetic, of which from an eighth to a quarter of a grain may be given in solution every two, three, or four hours, according to the urgency of the case.

If, in spite of these measures, the symptoms continue unabated, or the aperture of the glottis become narrowed by œdematous effusion, the doom of the patient is no longer doubtful. I have, in several instances, suggested for the consideration of eminent surgical authorities the propriety of performing tracheotomy; but on weighing the balance of probabilities, and calmly considering the dangers to be apprehended from an operation, even under favourable circumstances hazardous, it has been abandoned.

In a large proportion of cases, when enteric fever is well marked, the lungs are involved more or less seriously. The pulmonary affections are more common in the winter and spring, and often add greatly to the danger of the patient.

If there be merely catarrh, it generally subsides in a few days, yielding apparently with the other symptoms, without special treatment. But it may pass into a more serious affection—bronchitis— which is easily recognised by the irritating cough and local signs. If it be confined to the larger

tubes, a sinapism or the acetous solution of cantharidin, applied under each clavicle, with a saline
julep containing ipecacuanha wine and tincture of
henbane or syrup of poppy, will generally be sufficient to remove it. But when the inflammation
spreads to the smaller tubes, there is more difficulty
in dealing with it, and more especially when, from
its latency, it has not attracted notice at the commencement. If the powers are adequate to the loss
of a few ounces of blood from the chest, great relief
is generally obtained; but this can only be resorted
to when the general condition of the patient shows
constitutional vigour. In doubtful cases, or when
local bleeding cannot with safety be had recourse to,
excellent effects often follow the process of dry-cup-
ping. The application of a blister to some part of
the chest is indispensable; and if the bronchitic
affection has extended to both lungs, one should
be placed on either side of the chest. The antimonial saline julep, or a pill containing half a grain
of ipecacuanha with two of James's powder, and the
like quantity of extract of hemlock or poppy, at
intervals of four or six hours, are well suited, in
addition to the local measures.

The bronchitic affection, when severe, often
spreads to the air-cells, giving rise to pulmonary
congestion, which passes sometimes rapidly into
pneumonia, when it may be distinguished by its
physical signs. It is commonly confined to the
base of one lung, but may involve both, and,

according to its intensity, endanger the patient's safety. The treatment must be regulated not only by the extent, but by the duration of the affection; for it should be observed, that we seldom meet with secondary pulmonary disease until the more advanced period of the fever, or during convalescence. But unless there be special circumstances to forbid it, we must treat this secondary pneumonia on the same principles as the idiopathic, bearing in mind that we have to deal with it under very different circumstances. The measures embrace local abstraction of blood, or dry cupping, turpentine fomentation or blistering, and the exhibition of small doses of tartar emetic in combination with compound tincture of camphor, giving at the same time nourishment and wine if necessary.

When peritoneal inflammation arises in the progress of enteric fever, it originates from two causes—1st, from gradual extension of the ulcerative process in Peyer's patches until the peritoneum is reached, and subsequent inflammatory action in this membrane set up; 2ndly, from perforation of the intestine from the gradual process of molecular destruction. When it arises under the circumstances first mentioned, the management differs little from that of peritonitis occurring idiopathically, and therefore recourse must be had to the application of leeches to the abdomen, and warm fomentations, after which a pill, containing mercury and chalk with Dover's powder, is to be given at intervals of three or four hours, supporting the strength at the same time according to circumstances.

The treatment of peritonitis resulting from intestinal perforation can be considered palliative only; for though I have witnessed the recovery of two cases, in which the distinctive signs of this lesion were unequivocal, the termination is almost invariably fatal.

It is true that in some instances, as formerly mentioned, nature makes an attempt to repair the injury and avert the fatal issue, adhesion of the intestine at the point of perforation to some adjacent portion of bowel or organ taking place, so that the effusion of the alimentary contents is prevented. But this is a rare event.

Beware in the management of this formidable lesion of the abstraction of blood by leeches, which, while it is obviously inefficacious in controlling the inflammation, tends surely to exhaust the powers. The suggestions of Dr. Stokes are full of practical value. He states that the first indication of treatment is to support the strength of the patient as far as this can be done without injury; the second, to prevent the further effusion in the peritonial cavity, by endeavouring to induce organization and adhesions of the effused lymph. This latter indication is best fulfilled by time, and by attempting to diminish, as far as possible, the peristaltic motion of the intestines. For this purpose, opium is to be given in repeated doses. One grain of solid opium, or an equivalent quantity of laudanum, or Battley's preparation, is to be administered every second hour, till the symptoms of abdominal inflammation abate, after which, the same amount is to be repeated at more distant intervals. In one case, though unsuccessful, this treatment afforded decided relief. Sixty drops of the black drop were given in twenty-four hours. In another, in which recovery took place, one hundred and five grains of solid opium (exclusive of anodyne injections) were administered without the patient experiencing any of the usual effects of the remedy.

The form in which the opium is prescribed is, however, a matter of less moment than a knowledge of the principles on which it is administered. I have found a full dose—two grains of solid opium at once, followed by grain doses every second hour—answer every purpose; at other times I have given a third or half a grain of hydrochlorate of morphia in solution, and repeated according to circumstances with equally good results. Sometimes I have seen great comfort from tranquillizing the nervous system by small doses of laudanum, or the solution of morphia, at short intervals. The local administration of opium should not be overlooked, twenty drops of laudanum in three ounces of any mucilaginous fluid being from time to time exhibited by enema.

The strength of the patient must be well supported by strong broths, brandy and egg, animal jelly, and such like. If this nutriment be rejected by vomiting, as is not unfrequently the case, small quantities of strong soup or brandy and egg, with a few drops of laudanum, must be injected into the rectum.

TREATMENT OF TYPHUS FEVER.

Former treatment by bloodletting—emetics—purgatives—cold affu-
sion—salines and antimonials—nourishment—wine—revival of
the Brownonian treatment—narcotics—management of conva-
lescence.—Treatment of relapsing fever.—Treatment of febricula.

The pathology of typhus differs in many important
respects from that of enteric fever—in the absence
more particularly of special local lesions. It is,
moreover, a much less acute or sthenic disease
than its prototype (enteric fever), and consequently
the treatment, though in some respects similar,
differs, as we shall point out, in essential particulars. It appears, too, from the history of epidemics, that it has a much greater tendency to
spontaneous cure—a fact which explains the recovery of such a large proportion of those attacked
with the Irish famine fever without the aid of
remedies.

When the symptoms are mild, very simple measures are sufficient for its management; confinement
to bed, attention to ventilation, cooling drinks, mild
aperients, and suitable nourishment being all that
is necessary.

But by far the larger proportion of cases is of a
more severe character, requiring consequent modification in the treatment. Within my recollection
it was considered an indispensable part of the early

treatment of typhus to adopt a somewhat vigorous antiphlogistic plan; and with the view of subduing the general excitement, it was commenced by the abstraction of blood—an operation seldom omitted except under special circumstances—not reserved for particular cases, but adopted as a general rule, too often regardless of the stage of the disease or the condition of the patient. Of late years, however, this practice has been almost entirely abandoned, and in the present day is scarcely thought of as a remedy in maculated typhus. I have not ventured to abstract blood in a single instance, either in hospital or private practice, for many years past; but I can conceive that an isolated case may occur in which there is, in the early stage of the disease, so much constitutional disturbance, or the patient so plethoric, as to justify the loss of a few ounces of blood, in order to anticipate threatened congestion in some important organ. Such cases, however, are rare and exceptional. We may, in short, consider the abstraction of blood in genuine typhus to be now almost entirely proscribed.

The same views apply to the local abstraction of blood, when there is congestion in particular organs; but although blood is not to be taken, even locally, without due consideration, and unless there be evident necessity, it may, on the other hand, at times be imperative that the first threatening of undue local excitement should be checked.

In doubtful cases, or when loss of blood cannot be borne, dry cupping often affords much relief, without depressing the vital powers; and this practice is especially applicable when the patient is advanced in years, or when the powers are feeble.

Of evacuant remedies occasionally resorted to, I may allude to the exhibition of emetics, not with the object of recommending them, but simply to observe that, except under special circumstances, they are in the present day seldom resorted to. But if, in the early stage of typhus, there be indications of bilious derangement, such as sickness or vomiting, epigastric uneasiness, yellow coating of the tongue, with jaundiced conjunctiva, an emetic may be prescribed,—the act of vomiting being often followed by manifest improvement in the symptoms and great relief to the feelings of the patient.

Thirty years ago the employment of purgatives formed an indispensable element in the treatment of continued fever, without regard to form or type; and certainly, if we consider the freedom with which they were prescribed after the publication of the work of Dr. Hamilton, of Edinburgh, with good effects, enteric fever must, as is generally admitted, have been rare at that time, if not altogether unknown. While I acted as clinical assistant to Dr. Hamilton in the Edinburgh Infirmary, I witnessed his practice, which was, on the whole, successful;

but some years afterwards, when the doctrines of Broussais were promulgated, the purgative system, though not abandoned, was much less freely adopted. It is also very probable that the type of continued fever being at that time more acute, the powers were more able to sustain the repeated exhibition of purgative medicines. But in the present day their employment in typhus is restricted to the commencement of the disease, to clear the bowels of accumulated or acrid contents, while in its subsequent progress, the occasional use of a mild aperient is all that is called for. Or, if there be unusual excitement, which it is deemed necessary to repress, the exhibition of one or two active purgatives may be advantageously adopted.

I have already noticed the practice of cold affusion in the treatment of fever, as recommended by the late Dr. Currie, and pointed out objections to its employment. It is now entirely superseded by the more safe plan of cold or tepid sponging, which should always, except in cases of extreme exhaustion, be effectually applied every morning, or more frequently if there be undue heat of the skin. It not only abates the morbid heat, but affords great comfort to the patient, besides cleansing the skin— a matter of no small importance, especially in the nursing ministrations of the poorer classes. When the head is hot, and the face flushed, the topical application of cold will often alone anticipate or subdue cerebral excitement; and in all such cases,

S

it forms an important and indispensable part of the treatment.

In regard to the class of remedies included under salines, they may be employed, not on the ground that they are beneficial, but that they are a harmless means of appearing to do something; for no intelligent practitioner places the slightest confidence in their utility. On this subject the late Dr. Graves very pertinently observed, "that in the treatment of fever, it is frequently of importance to gain time, and periods will occur in every long fever, in which there may be no direct indication for the exhibition of any powerful remedy; at the same time, such is the ignorance of non-medical persons, and the anxiety of the patient's friends is so intense, that they cannot imagine how it is possible for an attentive physician to let twelve hours pass away without doing something. The mere circumstance of seeing the fever going on, is sufficient proof to them of the necessity of making renewed efforts for its removal. This, however, is very excusable, and we could not treat fever in a satisfactory manner without medicines of an expectant character, and calculated to fill up the spaces intervening between those periods when active treatment is necessary. You are not to suppose that in ordering such medicines you are acting a dishonest part, and practising a deception unworthy of your profession; on the contrary, your conduct is perfectly just and proper, and though

you are convinced that no medicine is required,
still it will be necessary to prescribe something, if
you do not wish to lose the confidence of the patient
and his friends. It may be said that these are
mere prejudices, and above the dignity of a man of
firm and consistent character; but since prejudices
are intimately blended with human nature, and
constitute, as it were, a part of it, it is much better
in many cases to submit to them, particularly when
compliance does not involve a sacrifice of principle."

With this view, an alkaline citrate may be prescribed, either alone in the form of julep, or if
the febrile excitement be such as to require
repression, a sixth or an eighth of a grain of
tartar emetic may be added; or a pill, containing
two or three grains of James's powder (prepared
by Newbery), taken at intervals, according to circumstances. In some cases, in which the antimonials are less indicated, it may be sufficient that
the alkaline citrate be taken in the pyrexial state,
adding the antimonial during the paroxysm only.
But antimonial preparations are not to be indiscriminately employed; like other remedies of po-
tency, they should be reserved for special emergency
only, and even in such cases they should be given
with caution in maculated typhus, and their effects
carefully watched, so that on the first approach of
exhaustion they may be at once withdrawn.

In my observations on the treatment of enteric
fever, you may remember that I dwelt on the im-

portance of the proper regulation of nourishment from the very commencement; and certainly in maculated typhus, it is even more essential that the waste of tissue should be compensated by due and regular supply of properly selected food. In addition to the ordinary kinds of farinaceous aliment, good beef or chicken tea, in divided portions, should be allowed, and instructions given to the nurse that in the night the patient be fed, once at least in four hours, with beef-tea alternately with some farinaceous preparation, such as barley-water, groat gruel, or arrowroot, pleasantly flavoured. It is well to feel our way with the animal broth at first, by giving it in small quantities—two or three tablespoonsful—at a time, and if it disagree, or the patient complain of weight in the epigastrium, sickness, or increased feverishness, it should be withdrawn, and farinaceous food substituted for a day or two, when it may be again tried, and will probably suit better.

The medical attendant seldom escapes the importunity of friends to permit fruits to be given. It is better to be cautious in giving consent, for such articles do not always agree, and as they are unnecessary, it is better to forbid their use, unless in the form of subacid drinks, such as apple or tamarind tea, a solution of preserved jelly in water, orangeade or lemonade. While on the subject of drinks in fever, I may allude to the necessity of restricting the patient as much as possible from over-distending the stomach by copious draughts of fluids—a practice which not only fails to check thirst, but often brings on gastric disturbance. The sensation of thirst is, as formerly stated, almost entirely confined to the fauces and upper part of the pharynx, and is as much relieved by a small quantity swallowed slowly as by a larger gulped down at once.

In the progress of typhus, generally in the second week, but sometimes earlier, symptoms of exhaustion become evident, and indicate the necessity for stimulants; hence wine, or some form of alcoholic fluid, becomes one of the most important remedies in the treatment. The precise period, however, when they should be given must be regulated, not so much by the duration or stage of the fever, as by individual circumstances. Symptoms of exhaustion or sudden collapse, for example, may supervene at any period, even at the very commencement, and require stimulants to be promptly administered, and in quantities which can only be regulated by their effects. The state of the circulation, as indicated by the pulse at the wrist, but still more surely by the sounds of the heart, is the usual practical index to the determination of the question. In a former lecture I drew your attention to an excellent paper on this subject by Dr. Stokes, in which it is shown that the cardiac sounds form the only certain guide for the proper administration of wine in fever. When, therefore, we find the first or systolic sound so weak and

abrupt as scarcely to be distinguished from the
second (in extreme cases it is nearly annihilated),
and the pulse at the wrist beating in sympathy with
the heart, soft, compressible, and rapid, there must
be no hesitation in at once administering powerful
stimulants, the effect on the heart's action being
the guide as to the amount of either wine or brandy,
while medicinal stimuli—ammonia, chloric ether,
or other remedies of this class—are given in the
intervals.   It is especially necessary in such cases,
to enjoin that in the night, when the depression
is generally the most marked, nourishment and
wine be given steadily, and not to admit the too
common excuse offered by the nurse for withholding
them—that the patient was in such a sound sleep
that it was not deemed prudent to interrupt it.
Many a patient has thus been allowed to pass into
hopeless collapse, when, in all probability, by constant judicious stimulation, recovery might have
taken place.

It is always necessary to watch the effects of the
first few doses of wine.   If the pulse abate in frequency, become soft and more full, the tongue moist,
and the heat of the skin not increased; or, when
there has been delirium, if the patient become
more calm, and have intervals of sleep, we may feel
sure that the wine is doing good.   On the other
hand, if the pulse increase in frequency and
strength, the skin become hotter, and the patient
restless, flushed and excited, with throbbing of the

temporal and carotid arteries, we may conclude, either that wine is not suited to the case, or has been given too early, and should therefore be withdrawn. But, as a general rule, it is perhaps better to give wine a little too early than a little too late, since, if it appear to disagree, it is easy to suspend its use, though it may be very difficult to restore the vital powers if they have been allowed to remain too long unsupported.

Nor should the wine or brandy be discontinued until convalescence is fairly established; but as the symptoms for which the stimulants have been prescribed disappear, the quantity should be gradually abridged, by giving smaller portions and at more distant intervals.

In regard to the amount of wine and alcoholic stimulants that may be administered in typhus, no precise rules can be laid down, as the ever-varying circumstances presented by individual cases can alone determine this. It is prudent to begin with half an ounce or an ounce, and to repeat this amount at longer or shorter intervals, according to the effect produced. From six to twelve ounces may be considered to be an average daily allowance, but sometimes it is necessary to give two or three pints, or even more, in twenty-four hours, and, it is surprising to observe, that such quantity is not followed by the slightest intoxicating effect, even when the patient has been previously unaccustomed to stimulants. Indeed, in low fevers, the exhausted

state of the nervous system appears to be an anti-
dote to the effects of stimulants—in short, to create
a tolerance of wine and diffusible stimulants.

The wine should always be conjoined with
nourishment, in order to assist its due assimilation,
though in many cases the digestive powers are so
feeble that they are unable to elaborate even the
lightest articles of food, and therefore the wine or
brandy may be given simply diluted with water.

I have just alluded to the daily quantity of wine
that it may be necessary to prescribe in typhus,
and stated that no precise rules can be laid down,
as the circumstances of each case must determine
this. You are doubtless aware that there is a great
tendency in the present day to revive the Brownonian system, which flourished for a time in the
latter part of the last century, not only in fevers,
but in all acute diseases, without regard to individual peculiarities. The doctrine inculcated by
some teachers with respect to inflammation is, that
this process being a deranged nutrition, involving
supply and waste, and the waste being considerable
while the inflammatory process lasts, there must be
a compensating supply; that as the supplies for
the formation of the abnormal products of pus and
lymph must be drawn from the blood, or from the
tissues, or from both, the vital powers become exhausted in proportion to the organic disintegration
that takes place. Hence it is concluded, that the
more the inflammatory process draws upon the

blood, the greater will be the exhaustion of vital force, and the consequent effect upon the whole frame.

Upon this physiological theory of the phenomena of inflammation is based the overthrow of established therapeutic principles, on which its treatment has been for ages conducted. But surely even the abettors of this theoretical view must admit, that the object of treatment is to anticipate or prevent those so-called destructive processes; in other words, to promote resolution by all available means. Is this to be accomplished by extravagant doses of wine and brandy, regardless of the ever-varying condition of the system or period of the disease ?

Similar reasoning is adduced in regard to the phenomena and treatment of fevers, whatever be their type or special circumstances. It is against the indiscriminate employment of stimulants in fever that we protest, being convinced, that their proper administration requires as much consideration as is generally bestowed on other measures employed as curative agents.

The enormous quantities of wine and brandy recommended in even the early stage of fevers, whatever be the form, the individual circumstances, or whether there be local affections present, have often surprised me, and inclined me to doubt the accuracy of the statements. I have certainly seen intercurrent inflammation materially aggravated by the injudicious stimulation adopted, and on more than one occasion all the ordinary characters of acute delirium tremens supervene, when the unlimited administration of brandy had been left to the discretion of a nurse, who fancied that she was only obeying instructions when she poured down dose after dose. There is surely no practical philosophy in such indiscriminate abuse of a really valuable remedy when given on rational principles; and I deem it the duty of every physician who is convinced of the dangerous tendency of the Brownonian doctrine applied indiscriminately in the treatment of diseases, acute as well as chronic, to express his opinion boldly and decidedly, that the young and inexperienced practitioner may be warned of the dangerous consequences of this recently revived doctrine.

In regard to the selection of wines, I may observe that the stronger white kinds are generally to be preferred. Sherry, properly diluted, is on the whole best adapted, as being in common use as an article of diet among the better class of the community. Madeira also, and port, if sound and aged, are either of them well suited, but the quality should be above suspicion. Champagne is generally very grateful, and may be allowed as an auxiliary, but it should not be relied on alone, unless the condition of the patient is such as to require only a slight occasional stimulus, or when the stronger wines disagree. The same may be said of claret and the other lighter wines, any of which, if sound, may be selected, according to the fancy of the patient; but dependence should alone be placed on sound sherry, Madeira, or port, as the therapeutic agents.

In hospital practice, the less expensive wines are for obvious reasons selected. African sherry (from the Cape of Good Hope) is almost universally employed, from its cheapness and the proportional amount of alcohol it contains.

Of the more diffusible alcoholic fluids, brandy, whisky, or gin are in common use, and are preferable, when a more speedy and permanent stimulant is indicated. They may also be given in addition to wine in special cases.

Fermented liquors are sometimes employed, but they are more suited to the stage of convalescence. When, however, the patient has a desire for them, from having been previously accustomed to their use as an article of diet, they may be allowed either as a substitute for or in addition to wine or alcoholic fluids.

And this brings me to allude to the administration of yeast as a remedy in the treatment of typhus fever. It has been thought useful in cases accompanied with symptoms of putrescency, in addition to wine and other cordials. Dr. Stokes, who has used it extensively, speaks highly of its efficacy, and thinks it well suited to cases with purple extremities or gangrenous sloughing, if the stomach bear it. It is in general easily taken alone, or with such other remedies as may be indicated, and it has the additional advantage of being moderately aperient.

The best mode of administering yeast is to give two tablespoonsful in water, or in an equal quantity of camphor mixture, every three hours; or, should the stomach be irritable, three ounces of yeast with two of arrowroot or starch gruel, and five minims of laudanum, may be thrown into the lower bowel, and repeated at intervals of four hours.

From the trials I have made of yeast, I am satisfied that it is a remedy well suited to this low form of typhus fever, and deserving of more general employment.

It is necessary in many cases to calm the nervous system; and this brings me to the consideration of the exhibition of narcotics, from the judicious employment of which so much benefit is occasionally derived.

Wakefulness, or want of sleep, so common in typhus, requires the employment of opiates; for nothing is more exhausting than two or three nights passed without sleep. Of the different remedies of this class, experience has led me to give the preference to opium, or its alkaloids; but I generally succeed with a much smaller dose than is usually prescribed,—with, perhaps, ten or twelve drops of Battley's laudanum, or of the liquor morphiæ,—repeating the same quantity every two or three hours until sleep is procured. But we ought to proceed on sure grounds, and if there be undue vascular excitement, indicated by headache, acute delirium, flushing and conjunctival injection, the narcotic must be postponed until those symptoms are subdued. I have recently alluded to the beneficial effects of a combination of tartar emetic and laudanum in the cerebral affections of enteric fever, and need not do more, in the present instance, than state, that it answers equally well under the same circumstances in typhus, administering at the same time in the severer forms, which so closely resemble delirium tremens, wine or brandy, in such quantities and at such intervals as the case requires.

Though opium generally acts better than any other of the narcotic class, it disagrees with some patients; and when such idiosyncrasy occurs, we must give a trial to some other remedy, such as henbane, extract of poppy, or Hoffmann's anodyne liquor. Some speak highly in favour of the extract of belladonna as a narcotic when the pupil is so contracted as to be thought to contra-indicate the employment of opium; but I have no experience of it. I have often observed, however, that when patients, or those around them, have intimated that opium disagrees, if given by enema, it acts like a charm, in soothing the nervous system and promoting sleep, concealing, of course, that opium is administered. Fifteen or twenty drops of laudanum, added to three or four ounces of any bland fluid, is often sufficient, and is, in many cases, a preferable mode of administration.

As to the repetition of opiates, I would observe, that when the purpose for which they have been given has been accomplished, the remedy need not be repeated, unless the same circumstances recur. It should, however, always be at hand to be administered if necessary.

In no disease are the beneficial effects of blisters more satisfactory than in these severe cerebral affections of typhus fever. But we must be careful not to apply them if there be congestion. When this is subdued, the blister may be applied to the forehead and temples, which is preferable to the nape of the neck, in which situation vesication is often attended with much discomfort from the constant friction to which the surface is exposed.

In the pulmonary affections to which I shortly adverted in a former lecture,—whether they assume the form of bronchitis, pneumonia, or, more rarely, of pleurisy,—special local treatment becomes necessary. But it must be remembered that not only is the local malady itself of a low or asthenic character, but the general or constitutional powers are depressed, and consequently, active remedies not only fail in their object, but tend further to diminish the strength. Counter-irritation of the chest by turpentine stupes, or sinapisms, or the application of a blister, often gives marked relief.

The secretion of the pulmonary membrane should at the same time be promoted by antimony or ipecacuanha, while the general powers are sustained by wine and nourishment.

The management of the convalescent stage of typhus, as well as of other forms of fever, is not unimportant, though often neglected. Much benefit is derived from various remedies of the tonic class—quinine especially, or some bitter infusion, such as that of cascarilla, calumba, or gentian, to either of which the compound spirit of ammonia may be usefully added. The diet should be light and nutritious, and suited to the digestive powers of the patient, and it is always proper to make careful inquiry into the previous state of the assimilating powers, in order that a suitable dietetic plan may be laid down. No general rules can be given, as each case must be regulated by its own peculiarities, and if due care be observed in eliciting these, there will be seldom great difficulty in adapting the details.

TREATMENT OF RELAPSING OR RECURRENT FEVER.

When I described the characteristic features of this comparatively rare and somewhat singular form of fever, I mentioned that it occasionally constituted the character of an entire epidemic It more commonly appears as an intercurrent disease during the prevalence of an epidemic of either enteric or typhus fever.   It requires, in some respects, special treatment.

In the epidemic which appeared in Edinburgh in 1817 and 1818, and of which a description was given by Dr. Welsh, he expressed great confidence in the efficacy of copious bloodletting, and his apparent success has furnished the chief argument employed by those who support the change-of-type theory in explanation of the altered views of the treatment of fevers.   I had the opportunity of witnessing this epidemic, and the bold treatment it received at the hands of Dr. Welsh, and though it appeared, on the whole, successful, I am satisfied that it was carried to a much greater extent than the symptoms required; indeed, I am almost disposed to think, that the issue would.have been equally successful had bloodletting not been practised at all, for succeeding practitioners have found that the loss of blood was not only unnecessary, but in subsequent visitations of the same fever could not be sustained.   Dr. Cormack, for example, in the description he has given of the epidemic relapsing fever which prevailed at Edinburgh in 1843, tells us that he was urged to adopt bloodletting by several medical friends, who had seen the disease treated, in 1817, by copious bleedings, and, as it appeared to them, with the effect of shortening its duration, preventing relapse, and mitigating the muscular and arthritic pain.   In the trials he made, however, he found that although in most, if not in all the instances in which blood-letting was resorted to, the pain in the head was either relieved or entirely removed for a time, and the pulse rendered more soft and slow, the beneficial changes could not be entirely ascribed to the remedy, since the same relief was observed to take place as suddenly and unequivocally, in patients in the same wards, and with the same symptoms, who were not bled at all. It was evident that few could bear the loss of more than ten ounces, and that others became giddy and faint after two or three only had been abstracted; and though some of the patients declared that they felt quite free from headache after the bleeding, it returned soon after with its former severity. Indeed, this symptom often ceased spon-taneously, or at all events, was more uniformly and effectually relieved after the full operation of a purgative, followed by the application of a cold lotion to the scalp, and immersing the feet in warm water. Even the local abstraction of blood appeared of doubtful efficacy.

In the cases that have come under my individual observation, I could perceive no indication for em-ploying venesection, but I trusted to moderately free action on the bowels in the young and vigorous, though caution, even in the employment of purga-tives, was necessary in patients who were advanced in years, as well as in those whose powers were feeble. The exhibition of one or two doses of calomel and rhubarb at night, followed by castor

T

oil or a neutral salt in the morning, so as to clear
out the bowels effectually, was generally followed
by relief of the symptoms, a gentle action being
afterwards maintained by mild laxatives. Free
action of the bowels having been procured, there
was little necessity for other remedies. Thirst
was appeased by any of the common beverages;
or if it were necessary to show that the patient
was not neglected, the saline alkaline julep, or a
mixture containing a few grains of bicarbonate of
potash in camphor water, was prescribed at intervals. The diet was restricted to farinaceous matters
and beef-tea; in short, the ordinary treatment of
mild fever was, in its routine details, adopted.

For the vomiting, so frequent in the course of
the disease,—often, indeed, an early symptom in
the primary fever, as well as in the relapse,—the
application of a sinapism or turpentine stupe to the
epigastrium, followed by a pill containing opium,
camphor, and extract of nux vomica, repeated at
intervals, or creasote or prussic acid, alone or in
combination, were resorted to. The opiate enema,
or the introduction of solid opium as a suppository,
often controlled this symptom.

In severe and persistent headache, accompanied
by pulsation in the temporal and carotid arteries,
and flushing of the face, it was necessary to abstract blood by cupping and employ cold applications to the head.

If bronchitis, pleurisy, or pneumonia supervened

with unimpaired powers, dry cupping, or the local abstraction of blood by cupping, and a saline anti- monial julep, or a pill containing James's powder, at suitable intervals, with turpentine fomentations to the affected side, comprised the measures employed in the pulmonary complications.

In the cases accompanied by jaundice and hepatic or splenic dulness, small doses of mercury were given, followed by the application of a blister or turpentine over the affected region. It was unne- cessary to push the mercury far, so that it was with- drawn on the first appearance of its action on the salivary glands, and the iodide of potash in a bitter infusion substituted.

Muscular and arthritic pain, when severe, often causes great irritability and want of sleep. This symptom was generally relieved by a combination of calomel, Dover's powder, and extract of colchicum, given at bedtime, and repeated, if necessary, in smaller doses during the day. In less severe cases, the iodide of potassium, especially if combined with a small quantity of colchicum, answered extremely well.

It is necessary however to keep in mind that when the symptoms assume an adynamic cha- racter, indicated by rapid, weak, compressible pulse, cool, clammy, jaundiced skin, bilious vomit- ing, delirium, followed by more or less deep coma, powerful stimulants, such as ammonia and chloric ether, and the liberal administration of wine and brandy, must be perseveringly given; while external stimulation is had recourse to, by means of blisters applied to the neighbourhood of the head, and over the region of the heart. The same measures must be resorted to if sudden collapse supervene, which not unfrequently happens in persons advanced in years, whose powers are inadequate to struggle against an acute disease in which so many organs essential to life are implicated. In short, the various anomalies must be treated on general principles, and in determining and carrying them out, much judgment and tact are often required.

When alluding to the nature of this form of fever,—the relapsing,—I stated my opinion that it had, in many respects, a great resemblance to the periodic class; and from the attempts that have been made to anticipate or prevent the secondary fever by giving antiperiodic remedies, this idea must have occurred to others. With this view, the cinchona bark, or its alkaloid quinine, has been given in the apyrexial period; but though I have prescribed it myself, and seen it administered by others, I cannot say that it has any apparent influence in preventing the relapse. It is no doubt useful as a general tonic, and should therefore form part of the treatment of the apyrexial stage; but certainly no confidence should be placed in it beyond its tonic properties. Arsenic—another remedy of the antiperiodic kind—has been tried with the same object, but has equally failed in arresting the secondary fever.

It would therefore appear that the analogy between the relapsing and periodic fevers is confined to a certain resemblance in some of its features only; for it appears neither to arise from the same causes or poison, nor to be influenced by those remedies which are almost invariably successful in the treatment of malarious fevers. It is equally clear that relapsing fever differs in many respects from either enteric or typhus fever, not only in symptoms and progress, but in its exciting causes or origin, as well as its comparatively small death-rate. It is therefore in every respect a disease *sui generis*, and interesting as an occasional, though by no means new, form of fever.

TREATMENT OF FEBRICULA.

This disease, which appears to include the milder forms of either enteric or typhus fever,—for I have seen no reason to consider it a distinct form of fever,—runs its course in a few days without special treatment beyond confinement to bed, regulation of the bowels and diet. But if these precautions be disregarded it is possible that a more formidable disease may be kindled—possibly an acute intercurrent disease, requiring appropriate management. Hence the necessity of enforcing care in any form of febrile ailment, however apparently mild.

SEQUELÆ OF CONTINUED FEVERS.

Relapse—wakefulness—mental imperfection— inflammation of the eye — pulmonary tuberculosis — bronchial catarrh — sweating— diarrhœa—pyæmia—neuralgia—anasarca—phlebitis— gangrene of the extremities—bed-sores.

HAVING considered the treatment applicable to the several forms of continued fever, I have shortly to draw attention to the management of its sequelæ; for although the cessation of the symptoms, or stage of convalescence, proceeds, in a large proportion of cases, uninterrupted by any untoward event until health is finally established, sometimes this onward course is interfered with by some unlooked-for event requiring special treatment.

A temporary recurrence of fever (relapse) may in many instances be traced either to overtaxing the strength too soon, exposure to cold, or to indiscretion in diet. In either case the remedy is simple and obvious, and the indisposition is seldom protracted beyond a day or two. If it last longer, the whole phenomena of the fever have either been renewed, or some intercurrent local affection has supervened. You may remember that I pointed out the tendency in enteric fever to inflammation of serous membranes; and it is no uncommon circumstance to trace the connexion between an apparent renewal of fever during the convalescent stage and lurking inflammation in some organ—very often of the chest—and must be treated accordingly. On the other hand, you must bear in mind the possibility of a sporadic case of true relapsing fever occurring in an epidemic of either enteric or typhus fever, though by attending to its peculiar diagnostic features there is little difficulty in detecting it.

Wakefulness we have seen to be a very common symptom in both forms of fever, and although the return of refreshing sleep is one of the surest signs of approaching convalescence, many patients continue restless and sleepless, even after every other symptom has disappeared. Nor is the sleeplessness apparently connected with antecedent cerebral affection, as it occurs as often in the mild as in the more severe cases. The convalescence being retarded by the wakefulness, it is necessary to overcome it; and this may sometimes be effected by a tablespoonful of brandy, or two or three of sherry, in warm water or arrow-root, or half a tumbler of Burton ale, at bed-time. If these do not succeed, there is no alternative but the exhibition of opiates, — the common laudanum, Battley's laudanum, morphia, or henbane. I have known it necessary to continue the opiate for weeks after the convalescence was in other respects confirmed; and let me observe, that when it is determined to discontinue the remedy, it is better not to apprise the patient, nor even the nurse, of the intention, but to withdraw the narcotic from the prescription, and to practise the harmless deception of administering to the mind only.

Mental imperfection, notwithstanding the frequent and often severe lesions of the brain in both forms of fever, is singularly rare. Mania or general incoherence is almost unknown as a sequence, and even should some slight impairment of intelligence remain after an unusually severe cerebral affection, it is, as far as my experience goes, temporary only, and in due time disappears. In some cases, after the general symptoms have vanished, a certain degree of mental dulness, or incapacity to fix attention on matters which used to interest may be perceptible, or perhaps the memory may be in some degree defective, but these are never permanent, and gradually disappear as the convalescence becomes established.

Nor is deafness, so often present at some period of the disease, persistent; for although the sense of hearing may be dull even for weeks after the stage of convalescence, it is ultimately, with very few exceptional cases, completely restored—differing in this respect from the deafness resulting from the destructive otitis following the exanthematous fevers, and which in general permanently destroys the sense of hearing. It may, however, be necessary, when the deafness is protracted, to employ local remedies, with the view of stimulating the auditory nerve, or, should slight effusion have taken place around it, of promoting its absorption. Any of the ordinary counter-irritants may be applied behind the affected ear, or to the nape of the neck, such as the compound camphor liniment, sinapisms, or the more powerful liquor lyttæ.

Inflammation of the eye occasionally supervenes. It is sometimes confined to the conjunctival membrane; but it is often of a more severe character, involving the internal structure of the eyeball. Mr. Hewson* has given the details of cases of ophthalmia succeeding to typhus fever, and it has also been described by Dr. Jacob† and Mr. Wallace,‡ as having occurred after the Dublin epidemic of 1826, and after the Glasgow fever of 1843, by Dr. Mackenzie. It appeared likewise in a few instances after the Edinburgh epidemic of the same year. I have seen it occasionally in the Fever Hospital, chiefly in children whose constitutional powers were much impaired by want, and living in crowded ill-aired dwellings. It seems to have occurred frequently in the Dublin epidemic of 1826-7, for Dr. Jacob states that he had met with seventy or eighty cases within the year, all being of the severer forms of internal inflammation of the eye. The ophthalmia comes on

---

* Observations on the History and Treatment of the Ophthalmia accompanying the Secondary Forms of Lues Venerea, pp. 109-114.

† Transactions of the Association of Fellows and Licentiates of the King and Queen's College of Physicians, vol. v.

‡ Medico-Chirurgical Transactions, vol. xiv., p. 286.

sometimes during the fever, but more frequently a few weeks after convalescence, or even much later, and is almost invariably confined to one eye. According to Dr. Jacob, the transparent parts are rendered more or less clouded and opaque, the circumference of the cornea assuming a whitish or grey appearance, presenting an opaque circle resembling the arcus senilis, and the anterior chamber of the eye appearing cloudy, probably from thickening of the membrane of the aqueous humour. The lens itself often becomes partially opaque, reflecting light falling obliquely upon it, and presenting an opaline amber colour; and in this way vision is apparently destroyed. The iris, too, becomes altered in colour, and loses its brilliancy, and though in some cases its motion is unaffected, in others it is sluggish, and the pupil more or less irregular. In some instances, pus is deposited in the anterior chamber. As may be anticipated from the severity of the inflammation, vision is more or less impaired, some patients being only able to distinguish light from darkness, though it has been remarked, that the imperfection of sight is not always in proportion to the extent of the inflammatory action, but often greatest in the mildest cases.

In the cases observed by Dr. Mackenzie in the Glasgow epidemic of 1843, the attack of ophthalmia occurred at periods varying from three to sixteen weeks from the commencement of the fever. The character of the affection appeared to him to be congestion followed by inflammation of the internal parts of the eye, and especially of the iris and retina. The capsule of the lens, lining membrane of the cornea, and the sclerotic, subsequently became involved, the conjunctiva being but slightly affected. These appearances were accompanied by severe pain in and around the eye, aggravated during the night, considerable lachrymation, and dimness of vision. In some instances, the inflammation was apparently limited to one or two textures only, as the lining membrane of the cornea and anterior crystalline capsule. In others, when the vital powers were much depressed, the cornea sloughed, the iris protruded, the humours escaped, and the organ was destroyed.

Now in regard to the treatment of those secondary forms of ophthalmia, Dr. Jacob assures us that it is not attended with much difficulty. Bleeding, locally or generally, in proportion to the urgency of the symptoms, blistering where there is much pain or intolerance of light, purgatives, antimonial medicines, and opiate stupes, are obvious means of relief. Dr. Mackenzie, too, thinks that bleeding can rarely with safety be dispensed with, the recovery being very slow and uncertain when it is neglected. Some cases in children he treated by leeching; but in adults, venesection was always necessary.

The treatment must, however, depend on the circumstances of each individual case. In many, the powers are so depressed that even local depletion cannot be sustained. Blisters applied to the temple and forehead, warm fomentations to the eye, excluding the light, keeping the eye clean by frequent sponging with tepid water, and applying the linimentum calcis to the tarsal borders two or three times a day, embrace the local treatment, while the general strength is supported by suitable nourishment, and even wine, according to circumstances. Dr. Jacob speaks in favour of the extract of belladonna applied round the affected eye, and with this he advises calomel and opium to be given until the necessary mercurial effect is produced. But I should hesitate, as a general rule, to exhibit mercury in these ophthalmic cases, and should anticipate better results from a mild sustaining plan, of which quinine forms a part, combined with soothing local treatment.

Pulmonary tuberculosis has been supposed to be one of the sequelæ of enteric fever; but notwithstanding the decided opinion expressed by some in favour of this view, pulmonary consumption, so far as my observation goes, cannot be regarded as a sequel of either enteric or typhus fever. From the frequency of this disease in Britain, it may naturally be expected that consumptive persons should be attacked with fever; and it is moreover possible that if, in its progress,

a pulmonary affection arise, the latent tuberculosis
may become developed and pulmonary consumption established. In this way only (as I formerly
observed) can phthisis be fairly considered to be a
consequence of fever of any form. It is a common
circumstance in examining the bodies of persons
who have died of fever, to find disseminated tuber-
cles in the lungs and perhaps in other organs, but
such appearances are in no way connected with the
antecedent fever, except as a coincidence.

When there has been a catarrhal or bronchitic
affection of the lung during the fever, cough and
expectoration often remain for some time after,
depending on the persistence of the pulmonary
disease in a subacute or chronic form. This generally subsides in time, but it should not be left
unnoticed, and it is in such cases that direct tonics
are specially useful—such as the various preparations of steel, quinine, or zinc—in checking the
bronchial congestion on which the cough and expectoration depend.

Empyema, as a termination of pleurisy supervening on enteric fever, has been already noticed.
It can scarcely be called one of the sequelæ, but as
it may be the only remaining lesion, and keep up
an irritative secondary fever, it may be regarded
as a consequence of the pre-existing fever. When
the amount of fluid is considerable, especially if it
be purulent, there is little chance of its becoming
absorbed, and it is almost hopeless to expect benefit

from remedies supposed to promote absorption—mercury, iodine, or blistering. There is certainly one resource—evacuating the accumulated fluid by tapping the chest. But with what prospect of success ?

Towards the conclusion of fever free perspiration often takes place, and has generally been thought to be critical; but when once established, it is apt to continue during convalescence, and retard recovery of strength. It generally comes on in the night or towards morning, and has a sour, disagreeable odour; and as it is not only uncomfortable to the patient's feelings, but proves exhausting, it should be checked. The nitrate of bismuth, oxide of zinc, quinine, or the mineral acids, generally control it. Sponging the skin with vinegar and water twice a day should also be employed, the patient being at the same time encouraged to leave his bed.

In the gastric system we occasionally observe indications that some part has suffered prominently in the course of the pre-existing fever. For example, after typhus, when the patient has previously been subject to dyspeptic irritation, the tone of the stomach is feeble, and requires that the diet be carefully regulated during convalescence. Even with due attention to dietetic management, the patient often complains of epigastric uneasiness after food, followed by flatulent distension and acidity. These symptoms may occur also after enteric fever under the same circumstance of previous tendency to dyspepsia; and you may remember that, in speaking of the lesions incidental to this disease, I stated that softening of the mucous membrane of the stomach was not uncommon, and that its existence might be suspected when there was unusual irritability of the organ, especially vomiting, at any period of the fever. The treatment of the milder forms of gastric disorder consists in allowing only such articles of food as the stomach can digest easily, regulation of the bowels, and the exhibition of light bitters, such as calumba or cascarilla, with the addition of small doses—five to ten minims—of the tincture of nux vomica two or three times a day. The severer forms, accompanied with vomiting, require the application of turpentine stupes, or a blister to the epigastrium, with the exhibition at intervals of ten minims of diluted nitro-muriatic acid, and a single drop of Scheele's acid, in any of the ordinary bitter infusions. Creasote, alone or combined with laudanum or hydrocyanic acid, often answers well; and I have seen excellent effects, in obstinate cases, from the nitrate of silver in minute doses.

It is not unusual after enteric fever to observe the evacuations from the bowels continue thin and unformed. This may arise either from a congested state of some portion of the mucous membrane of the intestine, or possibly from the intestinal ulceration not having completely cicatrized. But it is always important to attend to this symptom, which interferes with the recovery of strength, or should it depend on ulceration, may become ultimately of serious import. If it do not yield to strict regulation of diet and occasional opiates, the treatment formerly recommended for this special intestinal affection must be resumed.

The spleen sometimes continues enlarged after convalescence from either enteric or typhus fever, occupying a larger space than usual in the left hypochondrium. In general, however, it gradually regains its natural dimensions under the prolonged use of quinine.

The pathology or nature of pyæmia is indicated by its name—admixture of puriform fluid with the blood. It was first described as a distinct disease in 1838, by Tessier, under the title of purulent diathesis. It is also known under the name of purulent infection.

Pyæmia may arise as a primary affection—that is, independent of, or not preceded by any other disease (this is, however, rare); or, as in the greater proportion of cases, as the consequence of some acute specific disease, in which way it becomes one of the sequelæ of continued fevers. With its connexion with wounds or injuries of parts, or with uterine phlebitis, we are not at present concerned.

It is presented to our observation under different forms:—1, of disseminated abscesses in the sub-

cutaneous cellular tissue; 2, of deposit of puriform
fluid in serous cavities, especially in the larger
joints; 3, of deposit of a similar fluid in the substance of internal organs—the lungs, liver, and
spleen more especially. It occurs at all ages, often
in the most insidious manner, without warning of
its advent, in the form, perhaps, of one or more
purulent deposits in the cellular tissue, which are
often regarded as critical, and when of small size are
generally absorbed. Although such superficial collections of puriform fluid are seldom accompanied by
much constitutional disturbance, when deep seated
and extensive they induce an irritative secondary
fever, which persists until the matter is discharged, either spontaneously or by surgical interference. But when the fluid accumulates in the joints,
or, as sometimes happens, in the serous cavities
of the chest or abdomen, or their contained viscera,
the constitutional disturbance is severe. There is
prolonged shivering, followed by intense heat of skin
and flushing, acute headache, and perhaps, delirium,
rapid wiry pulse, intense thirst, and altered countenance. These symptoms are followed by acute
pain in one or more of the larger joints, resembling
rheumatism, to which succeed swelling and obscure
sense of fluctuation, while on different portions of
the skin erythematous patches sometimes appear.
As the local affection proceeds, the symptoms
of the accompanying fever are aggravated; the
skin assumes, in many cases, a sallow or even

icteroid hue, the pulse becomes weaker, the tongue more dry and brown, the teeth and lips covered with dark sordes, with evidently increasing prostration, restlessness and muttering delirium, or coma and subsultus. Such severe examples generally prove fatal before the end of the second week, often earlier; but the patient may linger for weeks, and ultimately die from the purulent poisoning.

In regard to the precise nature of the purulent infection, which bears, in some respects, a strong analogy to erysipelas, though there be no question as to its origin in the admixture with the blood of some morbid product having a close resemblance to pus, there is great difference of opinion as to the precise nature of the morbid matter, as well as the mode in which it gets into the circulation. Tessier believed it to be some modification of the organism characterized by a tendency to suppuration in the solids and coagulable fluids. On this theory Dr. Jenner remarks, that amid much that is pathologically erroneous, the doctrine appears to contain an important truth—viz., that in a certain number of cases the febrile disturbance is established before any local disease is set up, and consequently before any pus is formed, and by inference, that the abscesses are, in such cases, merely the effects of a special alteration of the element from which that blastema is exuded out of which they are developed. (*Med. Times and Gazette*, 1853).

Sédillot affirms that the circulation of pus-corpuscles with the blood is the sole cause of pyæmia on the following grounds: 1, the invariable pre-existence of a centre of suppurative action; 2, the relation observed between the formation of pus in the veins, the passage of that liquid into the blood, and the development of pyæmia; 3, the presence of pus in the blood, verified by observation; and 4, the results obtained by the injection of pus into the veins of animals. Dr. Jenner disputes these conclusions, and adduces two cases to prove that pyæmia may exist without pre-existing abscess or ulcer, and denies that there is evidence of the relation between the formation of pus in the veins, its passage into the blood, and the development of purulent infection, or that pus can be detected in the blood; and as to the effects obtained by injecting pus into the veins of animals, he states that, as the symptoms and the situation of the disseminated abscesses are different in cases of purulent infection and artificially-induced pyæmia, it is improbable that the disseminated abscesses in the two have their origin in the same cause.

From these statements, it is evident that pathologists are not agreed either as to the origin and characters of the deleterious fluid, or the mode by which it becomes mixed with the blood.

In regard to the treatment of purulent infection, there is little to be done beyond combating urgent

symptoms and supporting the strength by nourishment, wine, and quinine. Opiates to soothe the
nervous excitement and procure sleep and relief
from pain are indispensable, and should be given
not at bed-time only, but at intervals during the
day; while opiate fomentations are applied to the
regions in which the purulent formations take
place.

At the termination of an attack of fever, the
patient sometimes complains of acute pain in the
lower extremities, often limited to the soles of
the feet, and evidently of a neuralgic character.
In some cases there is marked hyperæsthesia of the
cutaneous nerves, the slightest touch, such as
passing the finger over the skin, giving pain more
or less acute. In others, there is some degree of
numbness or slight loss of power in the lower limbs,
which alarms the patient while it lasts, but I have
never known it permanent. It is seldom necessary
to do anything in the way of treatment in either
of these affections, which are only temporary, and
disappear under the tonic treatment applicable to
the period of convalescence. If the numbness and
loss of power persist, small doses of the tincture or
extract of nux vomica may be prescribed, with
stimulating frictions of the affected part with the
horsehair-glove, or an embrocation containing tincture of arnica and ammonia.

Œdematous swelling of the lower extremities
occasionally appears, and seems generally to depend

on local weakness, the capillaries being unable to resist the gravitation of the blood when the erect posture is assumed. It is seldom necessary to interfere further than to place the limbs in a horizontal position until the strength is restored, and to give local support by a well-applied bandage.

It is well to be aware, however, that in some instances of anasarca consequent on fever, the urine, on being tested, exhibits albuminous impregnation, showing the anasarca, under such circumstances, to be connected with a congested condition of the kidney and atony of its secreting structure. The renal congestion and impoverished state of the blood permit the escape of serum with the urine, which consequently becomes albuminous, and it is in this way that the disappearance of the albumen under the use of tonics is explained. The albuminuria thus induced may become more or less obstinate, and the secreting apparatus of the kidney ultimately spoiled; and hence the possible origin of Bright's disease from fever, and the importance that it should be detected in its early stage, and receive appropriate treatment.

There is another form of swelling of the lower extremities which occasionally appears in the convalescent stage. It is a painful swelling of one limb, analogous to the phlegmasia dolens of puerperal women. It occurs both in males and females, and was first described by myself in the thirtieth volume of the Edinburgh Medical and Surgical

Journal. It is indicated by a sensation of stiffness in moving the affected extremity, which becomes uniformly enlarged and painful, especially at the inguinal region. The swelling has a white glistening appearance, without increased heat, and does not pit on pressure. It is possibly due to crural phlebitis, though I was at first inclined to ascribe it to inflammation of the areolar tissue, and this view was rendered probable by its terminating in two instances in diffuse suppuration; but as there has been no opportunity, so far as I know, of examining the condition of the limb after death, this point is undetermined. It often yields to rest and warm fomentation, but if there be local pain, the application of a few leeches to the groin is required. The horizontal posture should be enjoined so long as the swelling continues. In the chronic stage, a well-applied bandage from the toes upwards is useful.

In malignant typhus, one of the most formidable sequelæ is gangrene of the lower extremities. It is of rare occurrence—at all events in the wards of the Fever Hospital. I have seen one or more toes, or a part of the foot, or even the lower third of the leg, become suddenly livid, and pass rapidly into gangrene. The gangrene is generally preceded by acute pain in the part, the integuments assuming a dusky-red or pale-purple colour, on which vesications containing dark serous fluid often appear. In other instances there is no pain, and the first indication of gangrene is given by the livid or almost black hue of the part, which is cold and without the slightest sensibility.

It is said that gangrene may attack the upper extremities also, but this I have never observed.

The immediate cause of the gangrene is doubtless some interruption to the circulation, either from the heart being so feeble that it is unable to propel the blood to the extreme parts, or from mechanical obstruction in the circulation of the affected limb.

The indications of treatment are, to limit the extent of the gangrene, and to aid the natural efforts in the separation of the mortified parts. Presuming that the gangrene depends on an imperfect supply of blood to the part from either of the causes stated, we are to endeavour to stimulate the general and local circulation. The former is best effected by administering powerful stimulants internally—wine, brandy, ammonia, chloric ether; while the latter object is promoted by enveloping the affected parts in cotton wool steeped in port wine or camphorated spirit of wine, preventing evaporation by oil-silk. These applications are to be frequently renewed. When the gangrened parts show a tendency to separate, indicated by the formation of a line of demarcation, the parts should be enveloped in a warm poultice sprinkled with camphorated spirit of wine, and renewed every few hours. In all cases of this kind, opium should be given in doses of a grain (of solid opium) or a third of a grain of morphia, every four hours, or more frequently, if there be great pain and nervous exhaustion, withdrawing the remedy or giving it only occasionally when the nervous system is tranquillized.

Another form of gangrene, happily of rare occurrence, requires passing notice. I allude to the disease formerly described under the name of cancrum oris, but now distinguished by that of gangrenous stomatitis. It occurs in children only, the most common age being between four and seven, though it has been observed beyond this period. This affection is met with in the advanced stage of both typhus and enteric fevers (as well as in the exanthematous), and is characterized by the appearance of a circumscribed swelling of one cheek, the integuments having a red shining appearance, frequently involving also the upper and sometimes the lower lip, and accompanied with profuse discharge of saliva of peculiarly offensive gangrenous odour. On examining the internal or mucous membrane of the cheek, a foul looking ulcer corresponding with the external swelling is discovered spreading over the gums, and implicating in some cases the alveolar process of the jaw, so that the teeth become loosened and drop out. As the disease progresses, the swelling of the affected cheek enlarges, a harder and brighter red spot appears in the centre of the diseased mass, and ultimately a

dark circular point becomes visible, which gradually spreads both in extent and depth. Finally
the slough is detached, and perforation of the cheek
takes place, so that the affected structures become
blended in one extensive gangrenous excavation.
Though the sufferings of the child are often disproportionate to the extent of the local mischief, the
general powers are gradually undermined, and
death, preceded by coma, is the common result,
though the little sufferer has occasionally been
known to recover from this sad complication.

As to treatment, our efforts should be directed
towards arresting the progress of the gangrene.
The child should be well supported by strong soups,
jellies, and wine or brandy; and as to medicinal
measures, the chlorate of potash in combination
with quinine or other preparations of cinchona, administered in as frequent doses as the stomach will
bear. It is necessary also to administer now and
then small quantities of the mildest aperients, as
fluid magnesia, rhubarb, or castor oil, in order to
get rid of offensive and irritating secretions. In
regard to local applications, it has been recommended to cauterize freely the mortified tissues
by the free use of strong nitric or hydrochloric
acid. This is best applied by means of lint fastened
on a glass rod or a piece of wood or whalebone, the
acid being from time to time re-applied according
to circumstances. A single application may be
sufficient, but in general a repetition is necessary,

so as to destroy effectually the sloughs as they appear. After the use of the strong acid in this way, a lotion of chloride of lime or soda, or Condy's fluid, should be applied by a syringe now and then, while lint moistened with the same lotion is used as a dressing in the intervals.

Bed-sores are a very troublesome, and when extensive, dangerous result of low fevers. They generally interfere materially with the convalescence, from the drain on the patient's strength, by profuse suppuration and consequent irritative fever. I have often known them to be even the immediate cause of death. Though they seldom appear till the middle or latter stage of the fever, in corpulent subjects, or when from prostration the patient is unable to change his posture, the parts which are subjected to pressure may give way at an earlier period. Bed-sores form on the sacrum, trochanters, knees, ankles, or elbows, and even on the ears, scapulæ, or angles of the ribs. The skin covering the part first inflames, vesication or detachment of the cuticle succeeds, the cutis loses its vitality, and an eschar or superficial cutaneous gangrene of varying extent forms, and becomes slowly detached from the surrounding portion of healthy skin, leaving a more or less deep excavated ulcer of unhealthy aspect, sometimes even exposing a portion of carious bone.

In the treatment of bed-sores, a point of importance is, that with the object of removing

pressure from the inflamed or ulcerated parts, the
posture should be frequently changed. Great relief is often afforded by keeping the patient on his
face, but better by placing him on the water-bed
invented by Dr. Arnott, which is indeed indispensable in the management of bed-sores. I have
seen an extensive inflammation of the sacrum or
trochanters, which would have probably soon ended
in sloughing, disappear after the invalid was placed
on the water-bed, which has superseded many in-
ventions having the same object in view—removal
of pressure.

If there be merely erythematous inflammation of
the skin, the inflamed surface may be brushed over
with collodion, either alone, or mixed with castor
oil or cod-liver oil in equal proportions, or with a
solution of bichloride of mercury in proof spirit, in
the proportion of two grains to the ounce. I have
also seen in the early stage good effects from a
popular remedy—a solution of alum in any weak
spirit; but still better results from a solution of
nitrate of silver (ten grains to the ounce); or should
a solution not be at hand, the parts may be pencilled
with solid nitrate of silver. If these local measures
fail in preventing the destruction of the skin and
subjacent cellular tissue, the separation of the mortified integument should be promoted by poultices,
composed of boiled carrots, bread and water, or
linseed (sprinkled, if the odour be offensive, with
peat charcoal) and renewed three or four times a

day; or, what seems to answer well, by pledgets of lint soaked in warm water, and covered with oil-silk, which may be continued after the eschar has separated. If the ulceration begin to look pale and unhealthy, it will be necessary to use a gently-stimulating application. It may be dressed with any of the balsamic preparations, such as Peruvian balsam and castor oil, in the proportion of one of the former to two of the latter, applied on pledgets of lint. Another plan is to apply the warm-water dressing, and, in addition, to wash the bed-sore night and morning with a solution of chloride of sodium, in the proportion of twenty to thirty drops of the concentrated solution to an ounce of water, or with a weak solution of chloride of zinc. Some prefer the nitric acid lotion (made by adding two drops of the concentrated acid to a pint of water), or the common black wash; others, a weak solution of the nitrate of silver (five grains to the ounce), once a day, the sore being afterwards covered with carded cotton-wool and oil-silk. If there be great pain in the sores, the solution of morphia may be added to any of the applications preferred, as well as given internally.

The tannate of lead has been strongly recommended by M. Autenreith, of Vienna, as an excellent application to bed-sores. It is made by boiling an ounce of oak bark in eight ounces of water, and adding a solution of the diacetate of lead, (liq. plumbi diacet.) The precipitate (tannate of lead), is collected on a filter, and, after being washed, is applied to the bed-sores, which are afterwards to be covered with a poultice. This dressing is to be renewed night and morning. It may also be used in the form of ointment (equal parts of the tannate of lead and axunge) spread on lint, the sores being afterwards covered with cotton wool and oiled silk.

The cod-liver oil has also been proposed as an application to extensive bed-sores with caries of bones. Lint soaked in this oil is to be laid over the ulcer, and renewed night and morning.

If, in the process of healing, the granulations become exuberant, they may be touched with a concentrated solution of sulphate of copper.

THE END.